Chicago Price Theory

Chicago Price Theory

Sonia Jaffe, Robert Minton,
Casey B. Mulligan, and Kevin M. Murphy

Princeton University Press • Princeton and Oxford

Published by Princeton University Press
41 William Street, Princeton, New Jersey 08540
6 Oxford Street, Woodstock, Oxfordshire OX20 1TR
press.princeton.edu

Library of Congress Control Number: 2019942325
ISBN 978-0-691-19297-0
ISBN (ebook) 978-0-691-19881-1

British Library Cataloging-in-Publication Data is available
Editorial: Joe Jackson and Jacqueline Delaney
Jacket/Cover Design: Lorraine Doneker
Production: Erin Suydam
Publicity: Nathalie Levine and Julia Hall
Copyeditor: Alison S. Britton

This book has been composed in Sabon

Printed on acid-free paper. ∞
Printed in the United States of America

10 9 8 7 6 5 4 3 2 1

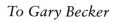

To Gary Becker

Contents

Acknowledgments   *xiii*

Chicago Price Theory: An Introduction   1

    The Chicago Economics Tradition   1
    Price Theory Differs from Microeconomics   2
    Using *Chicago Price Theory* to Learn Economics   4
    Example: Ethanol Fuel Subsidies   6
    Example: Acquired Comparative Advantage   12
    Outline of the Course   16

PART I: Prices and Substitution Effects

Chapter 1
Utility Maximization and Demand   21

    Utility Maximization   21
    The Theory of Demand   25

Chapter 2
Cost Minimization and Demand   30

    The Cost Function   30
    Hicks' Generalized Law of Demand   34

Relationships between Indifference Curves and the
   Demand System   35
Properties of Hicksian Demand Functions   36

Chapter 3
Relating the Marshallian and Hicksian Systems   38

The Slutsky Equation   38
Adding Up and Symmetry for the Marshallian System   41
Demand System Degrees of Freedom   43
The Income Effect of a Price Change   44

Chapter 4
Price Indices: Consumer Theory Guides Measurement   48

Laspeyres and Paasche Decompositions of
   Expenditure Growth   48
Chained Price Indices   51
Using the Cost Function to Value Quality Change   55

Chapter 5
Nudges in Consumer Theory   59

Indifference Curves for Buyers   59
Consumer Misinformation and "Nudgeability" Is a Prediction
   of Consumer Theory   60

Chapter 6
Short- and Long-Run Demand, with an Application to Addiction   62

An Example: The Demand for Cars and Gasoline   62
Relating the Short-Run Demand Curve to the Overall
   Demand System   64
Using Consumption Stocks to Understand Addiction   66
Short- and Long-Run Price Effects on Addictive Behaviors   69

Homework Problems for Part I:
Prices and Substitution Effects   73

PART II: Market Equilibrium

Chapter 7
Discrete Choice and Product Quality   79

Market Demand Is a Distribution Function   79
Equilibrium Product Quality   81
Heterogeneous Firms   87
Heterogeneous Firms and Consumers   91

Chapter 8
Location Choice: An Introduction to Equilibrium Compensating Differences   94

Properties of the Rent Gradient Model   96

Chapter 9
Learning by Doing and On-the-Job Investment   101

Human Capital Acquired from Training Programs Administered
   by the Employer   101
Learning by Doing   102
Types of Human Capital   104

Chapter 10
Production, Profits, and Factor Demand   106

Comparative Advantage and the Production-Possibility
   Frontier   106
The Production Function   109
Profit Maximization   110
Cost Minimization   112
The Firm's Slutsky Equation   114
Two-Input Production   116
Substitution and Scale Effects on Factor Demand   120
Acquired Comparative Advantage   121

Chapter 11
The Industry Model 126

 Properties of the Industry Model 126
 The Supply-Demand Perspective on Industry Behavior 128
 Four Ingredients of the Industry Model 131
 Industry Elasticity of Labor Demand 132
 Are Labor and Capital Complements or Substitutes? 133

Chapter 12
The Consequences of Prohibition 135

 The Revenue from Drug Sales 135
 The Legalization Multiplier 136
 Half-Hearted Prohibitions Are the Most Costly 137

Chapter 13
A Price-Theoretic Perspective on the Core 140

 Looking for Gains from Trade: Indifference Curves for Buyers
  and Sellers 140
 Exclusive Dealing, Quantity Discounts, and Other Market Outcomes
  That Are off the Marshallian Demand Curve 142

Chapter 14
Multiple-Factor Industry Model 145

 Review of the Industry Model 145
 Properties of the Multiple-Factor Industry Model 146
 Analyzing Production 148
 Endogenous Factor Prices 149

Homework Problems for Part II: Market Equilibrium 150

PART III: Technological Progress and Markets for Durable Goods

Chapter 15
Durable Production Factors 155

 Stocks and Flows for Factor Prices and Quantities 155
 The Use and Investment Markets for Capital Goods 157

Four Equilibrium Conditions   157
Steady State   159
Perturbing the Steady State   159

Chapter 16
Capital Accumulation in Continuous Time   166

Perturbing the Steady State (Continued)   166
Continuous-Time Versions of the Four Equilibrium Conditions   168

Chapter 17
Investment from a Planning Perspective   172

Adjustment Costs Applied to Net Investment   174
Endogenous Interest Rates: The Neoclassical Growth Model   176

Chapter 18
Applied Factor Supply and Demand 1: Technological Progress
   and Capital-Income Tax Incidence   180

Definitions of Labor Productivity   180
Explaining Economic Growth in the Presence of
   Complementarity   180
The Consequences of Unbiased Technological Change   182
The Incidence of a Capital-Income Tax   184
Why Capital is Elastically Supplied in the Long Run   186
The Incidence of a Corporate-Income Tax   186

Chapter 19
Applied Factor Supply and Demand 2: Factor-Biased
   Technological Progress, Factor Shares, and the
   Malthusian Economy   189

The Definition of Technological Bias   189
Relating Labor's Share to Economic Growth   191
The Malthusian Special Case   195
Capital-Biased Technical Change Also Benefits Labor   195
Adding Human Capital   197

Chapter 20
Investments in Health and the Value of a Statistical Life   199

   Investments in Self-Protection   200
   The Value of a Statistical Life   204

Homework Problems for Part III: Technological
Progress and Markets for Durable Goods   207

Notes   211
Bibliography   217
Index   221

Acknowledgments

In addition to Gary Becker, to whom we've dedicated this work, we would like to thank the many others who taught us price theory, both inside and outside the classroom: Robert Lucas, Sherwin Rosen, Jose Scheinkman, and Robert Topel at the University of Chicago; Armen Alchian, Ben Klein, Mike Ward, and Finis Welch at the University of California Los Angeles; and Robert Barro, Edward Glaeser, and Hendrick Houthakker at Harvard University. We also thank MarrGwen and Stuart Townsend for initiating and supporting this project.

Thanks as well to the many students of price theory, from whom we have also learned and whose interest and enthusiasm motivated us to create this text so they can more easily continue to teach and spread the approach and techniques of price theory. In-class questions from the University of Chicago graduate students in the entering classes of 2015, 2016, and 2017 contributed to the examples given in the text and videos. We also received valuable comments and encouragement on elements of this manuscript from Dora Costa, Michael Dinerstein, Joe Jackson, Matthew Kahn, David de Meza, Ging Cee Ng, Emily Oster, Tomas Philipson, Yona Rubinstein, Jesse Shapiro, Pietro Tebaldi, economists at the Council of Economic Advisers, and anonymous referees.

We are also grateful to Daniel Chavez and Virginia Bova, who assisted in the compilation and editing of this manuscript.

Chicago Price Theory

Chicago Price Theory

An Introduction

THE CHICAGO ECONOMICS TRADITION

A longstanding Chicago tradition treats economics as an empirical subject that measures, explains, and predicts how people behave. Price theory is the analytical toolkit that has been assembled over the years for the purpose of formulating the explanations and predictions, and guiding the measurement.

In the tradition of Chicago's "Economics 301," the purpose of this course is to help you master the tools in the kit so that you can use them to answer practical questions. Studying price theory at Chicago is "a process of immersion in those models so that they bec[o]me so intuitive to one's work that, in combination with new empirical investigation, they open the door to novel evaluations of market organization and government policy."[1]

Because price theory at Chicago has always been tethered to practical questions, this course and the course Jacob Viner taught at Chicago almost 90 years ago (Viner 1930/2013) share some remarkable similarities. The tradition draws heavily on Alfred Marshall (1890) in, among other things, viewing human behavior in the aggregate of an industry, region, or demographic group. Market analysis is essential to price theory because experience has shown that markets enable each person to do things far differently than if he or she lived in isolation. It is no accident that price theory is named after a fundamental market phenomenon: prices.

Price theory is not primarily concerned with individual behavior; models featuring individuals are provided when they offer insight about the aggregate. None of this is to say that price theory only looks at average or representative agents. Indeed, a primary reason that markets transform human activity is that they encourage the amplification of innate differences among people. Heterogeneity can be important; as we see in the example of comparative advantage below, markets can amplify heterogeneity through returns to specialization.

Price theory has not been static, though. Gary Becker, who taught Economics 301 for many years and gives a couple of the lectures in the video series that accompanies this book, developed human capital analysis and extended price theory to deal with discrimination, crime, the family, and other "noneconomic" behaviors. Becker and Murphy revisited the topic of complementary goods, using it to examine addictions, advertising, and social interactions (Becker 1957, 1968, 1993; Becker and Murphy 1988, 1993, 2003). Most important, people and businesses are in different circumstances today than in Viner's time—as witnessed by the decline of agricultural employment, increased life expectancy, and the rise of information technology.

PRICE THEORY DIFFERS FROM MICROECONOMICS

Although strategic behavior, such as the interactions among sellers in a market where they are few in number, has been treated with price theory (Weyl 2018), the introductory Chicago price theory course has not emphasized it. Competition, by which we mean that buyers and sellers take prices as given and the marginal entrant earns zero profit, is emphasized in large part because for most purposes, it is a reasonable description of most markets (Pashigian and Self 2007). Moreover, the competitive framework is simple enough to make room for us to master additional aspects of tastes and technology—such as product quality, habit formation, social interactions, durable production inputs, and complementarities—that are important for practical problems. Monopoly models are used on those occasions when price-setting behavior is relevant (Friedman 1966, 34–35; Stigler 1972; Demsetz 1993, 799). More generally, price theory is stingy as to the number of variables that are declared to be important in any given application.

In emphasizing markets and competition, price theory is different from microeconomics. Both typically begin with the consumer or household, but price theory stresses how consumers react to prices, many times without reference to utility or even "rationality"; whereas microeconomics takes care to lay down an axiomatic foundation of the utility function and individual demand functions. Price theory then quickly gets to market equilibrium, treating related subjects such as compensating differences, tax incidence, and price controls.

Microeconomics makes more intensive use of game theory, which traditionally puts somewhat more emphasis on rationality and optimizing agents. Both price and game theory model behavior as an equilibrium, but the latter typically focuses on interactions among small numbers of agents and strives to make separate predictions for each one. The rest of the market is treated as a constant.

The typical auction model of price (Klemperer 2004) is an example of the game-theoretic approach. That model has a fixed number of goods for sale in the auction, with little attention to how the goods were produced or how they would be used if not sold in the auction. The model has a fixed number of buyers and predicts how each buyer separately makes bids on the items for sale. Understanding why there are, say, two buyers rather than some other number, or what determines the seller's reservation price, is considered to be an advanced topic. With its emphasis on competitive market equilibrium, basic price theory is not concerned with bid prices but rather the ultimate transaction price, aggregate quantities produced and sold, and how they are connected with costs of various kinds, as well as how the good is situated in the consumer demand system.

The market-equilibrium approach says that the most important effects of policy, technical change, and other events are not necessarily found in the immediate proximity of the event. An ethanol subsidy example, discussed below, features a subsidy that is paid only in the market for fuel, which uses just a fraction of total corn production but has more price-sensitive demand. The market for animal feed is unsubsidized, but corn farmers' opportunity cost for selling animal feed is linked to the subsidized fuel market, so much of their gain from the subsidy comes from the increase in the equilibrium price of animal feed.

Real-life situations involve an element of strategic interaction where the players in a small-scale game understand the outside options available to them in a larger market. One approach would be to simultaneously model both the strategies and market prices. Auction models could, in principle, have endogenous production, entry, and reservation values that reflect economic activity outside the auction. But the point of theory in economics or any other field is to focus on important variables and leave the others to the side. As noted above, a great many markets have many buyers and many sellers, and have complementarities, taxes, habits, and other variables that need attention before getting into the strategic details for specific buyers or sellers. These are the situations in which price theory is needed.

The ethanol subsidy example also demonstrates how price theory guides measurement. Empirical studies of markets over time, or comparisons across countries or industries, must consider how to summarize a seemingly complicated reality behind each observation. Price theory shows how the appropriate approach to measurement depends on the question at hand.

Putting practical questions in a market context changes the answer. Trained economists are generally aware that market analysis is why the economic incidence of, say, a tax is different from the legal liability for paying the tax. But without price theory, economics training has too little practice in market analysis and results in policy investigations that too quickly presume that, say, the corporate income tax primarily harms corporations or an earned income tax credit primarily benefits workers.

USING *CHICAGO PRICE THEORY* TO LEARN ECONOMICS

Graduate microeconomics texts often devote more pages to game theory than competitive equilibrium, and part of their competitive analysis is dedicated to confirming that an equilibrium exists as a mathematical object. To the price theorist, the toolkit's mathematical foundations and possible abstract generalizations are an interesting subject for specialists, whereas a general economics education requires seeing how the tools have been successfully applied in the past and preparing to nimbly apply them to the next practical question that we encounter. Completing a mathematical microeconomics course will not make you good at price theory; price theory skills are obtained by practicing applications of the toolkit.

Whereas many economics courses help you master models, and leave application of those models as an advanced topic, price theory immediately engages the student with applications. The book and video series (available at https://press.princeton.edu/titles/30205.html or ChicagoPriceTheory.com) together provide three or four methods of practicing applications. First, both book and videos contain chapter-length examples such as addictive goods, urban property pricing, learning-by-doing, the consequences of prohibition, the value of a statistical life, and occupational choice. These chapters are instances of applications of price theory that were advanced by important research papers, and sometimes spawned an entire subfield of research activity, with novel and counterintuitive results.

At Chicago, both the students and instructors over the years have gotten better at price theory as a result of engaging with the homework. If you want a formula that makes you good at price theory, this is it: practice. Know what tools are available to study markets, and with repetition notice the types of questions to which each tool is best suited, in the sense of offering a simple analysis with predictions in accordance with observation.

The Chicago homework problems are not paired with specific lectures because part of excelling at real-world applications is knowing which price-theoretic tool is the best one to use for a particular practical problem. This book therefore provides a number of sample homework questions, but only at the end of one of the three parts of the book. The video series includes about a dozen of Professor Murphy's impromptu answers to student questions about current market events.

Becker and Murphy's course has always been intensive in solving applied problems, with considerable time of the instructors and advanced star graduate students devoted to formulating and helping students solve homework questions. The drafts of the book and video are now being used at Chicago to further "flip" the Price Theory classroom so that more of the student interactions with Murphy address applied problems.[2] Price theory instructors not at Chicago also have the opportunity to reallocate their time away from lecturing—let this book and video series help with that—and toward developing and discussing relevant and challenging applied homework questions.

Another way to practice applications is to do some homework before you begin the course and return to them afterwards. You will be amazed at how differently you think at the end! The six questions below are good examples:

1. Is learning by working on the job cheaper than formal schooling? (See chapter 9.)
2. What is the difference between prohibiting marijuana sales and subjecting its sales to a high tax? (See chapter 12.)

3. A great many manufacturers use machines and labor in fixed proportions. Does that mean that the wage rate has little effect on the amount of labor used in manufacturing? (See chapter 7.)
4. Does the availability of e-books reduce the sales of physical books? (See chapter 11.)
5. When housing prices are above their long-run values and continue to rise, is that good evidence that home buyers or builders have unrealistic expectations about the future? (See chapter 15.)
6. Could a billion dollars in federal subsidies to farmers increase farm incomes by more than one billion? (This chapter.)

As you work through the homework questions and the applied chapters, you will practice identifying and applying the tools of price theory. But the tools are just a means to an end, which is to understand human behavior. Most of the homework questions and applied chapters in price theory are therefore real-world questions about human behavior, of the same kind that are addressed by professional economists every day at central banks, major corporations like amazon.com, and regulatory agencies like the Food and Drug Administration.

Because it is useful, price theory gets applied to a large number of practical questions. Each practitioner of price theory thereby builds a wealth of experience that pays dividends in subsequent applications. New problems are recognized for their relation to problems already solved. Perhaps this is why price theory is sometimes called "intuitive."[3]

EXAMPLE: ETHANOL FUEL SUBSIDIES

A Market "Multiplier"

The federal government has been supporting the production of ethanol fuel with a variety of tax credits, subsidies, guarantees, and so forth. When the U.S. government started subsidizing ethanol fuel, the price of land used to grow corn—the primary ingredient in U.S. ethanol production—increased, regardless of whether the corn grown on that land actually ended up in the fuel.

Given that U.S. ethanol is primarily produced with corn, is it possible that corn farmers benefit by more than $1 billion for each $1 billion that the federal treasury spends on that support? In other words, let's use price theory to examine the incidence of ethanol fuel subsidies.

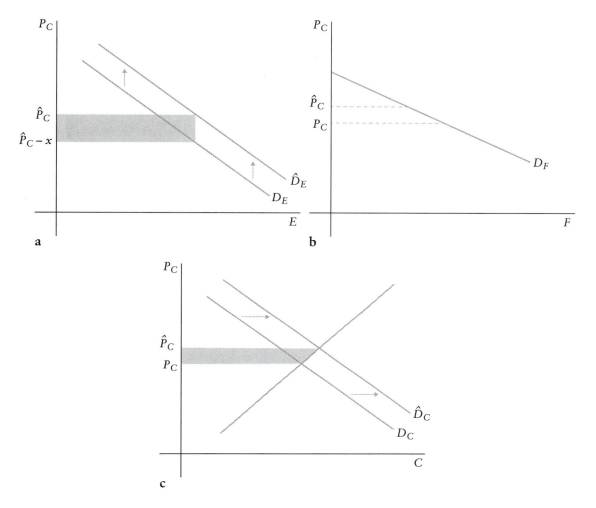

Figures I-1a, I-1b, and I-1c: Can farmers gain more from an ethanol subsidy than the amount the government pays?

Take a simple model in which corn, C, is used to make either ethanol fuel, E, or animal feed, F. We will consider demand curves D_E, D_F, and D_C, shown in Figures I-1a, I-1b, and I-1c, respectively; D_C, the market demand curve for corn, is found by adding the demands for ethanol and animal feed. A subsidy of the amount x per unit corn used in ethanol serves to increase the demand for ethanol by x units in the price dimension to $\hat{D}_E$. Horizontally adding the new ethanol demand curve with the stable feed demand curve, we get a new overall corn demand curve $\hat{D}_C$. Supply and demand for corn determine the equilibrium price of corn, which is the same regardless of how it is used. An example of our market is shown in Figures I-1a–c.

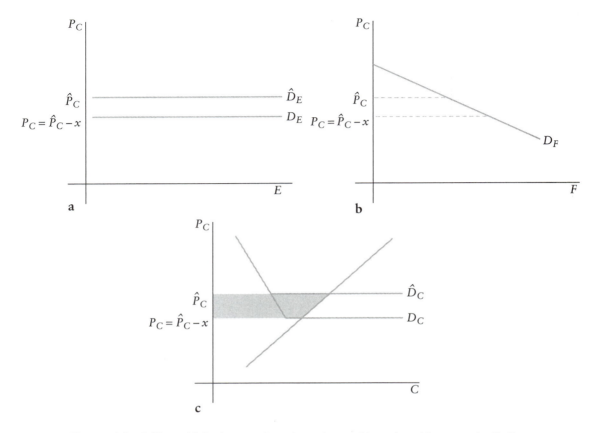

Figures I-2a, I-2b, and I-2c: In a market where demand for ethanol is more elastic than the demand for feed, the benefit of the ethanol subsidy to corn farmers can exceed the amount the government spends on the subsidy.

The result of the subsidy is that more corn is sold overall, and for a higher price ($\hat{P}_C$ rather than P_C). Less corn is sold for animal feed, because that demand curve is stable and the price is higher. The extra corn sales go to ethanol because the subsidy amount x more than offsets the price increase.

Our question, posed from the perspective of the figure, is whether the producer-surplus trapezoid in the market for corn (see Figure I-1c) can be larger than the subsidy-expenditure rectangle in the market for ethanol (see Figure I-1a).

Consider a case in which the demand for ethanol fuel is perfectly elastic (Figure I-2a) and the demand for feed is strictly decreasing (Figure I-2b). The overall demand curve is flat when the price is below what the ethanol market will bear (Figure I-2c). At prices above that, all corn is sold for

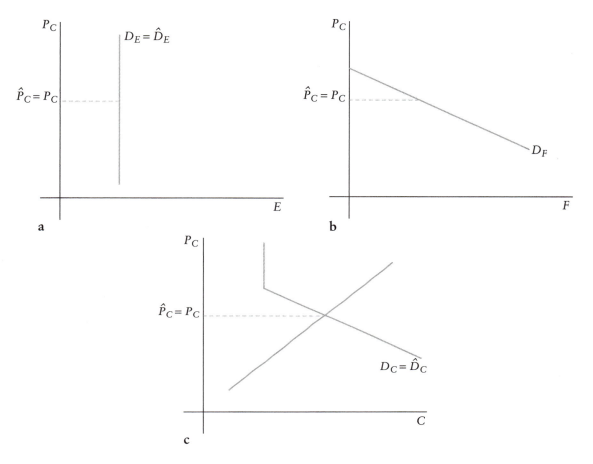

Figures I-3a, I-3b, and I-3c: In a market where demand for ethanol is more inelastic than the demand for feed, the benefit of the ethanol subsidy to farmers cannot exceed the amount the government spends on the subsidy. The ethanol demand shown above is perfectly inelastic, so the subsidy has no price impact.

animal feed and none for ethanol. Putting the two together, we have an overall demand curve with a hockey-stick shape, as shown below when we adapt the previous graphs to this new setting, as shown in Figures I-2a–c.

Suppose the subsidy is $0.10 per gallon. Then, in this market, the $0.10 gap created between the buyer and seller price per gallon in the ethanol market gets carried over in full to the aggregate market for corn.[4] If the subsidy is small enough, the gain to corn farmers is larger than the amount the government is paying.[5] Why? Not only do corn farmers get $0.10 more for the corn going to ethanol, which the government pays; they also get $0.10 more for the corn going to feed, which the animal-feed buyers pay. Maybe this also helps explain why the federal government assists corn farmers with an ethanol subsidy rather than paying the farmers cash directly.

Now consider a case in which the demand for ethanol fuel is perfectly inelastic. We leave the demand for feed unchanged.

Figure I-3a shows ethanol corn demand as perfectly inelastic, which means that, given any price, people demand the same amount. Thus an ethanol subsidy, which reduces the price that the ethanol corn buyers see, has no effect on their demand. Because the market demand curve is just the sum of the demand curves in the ethanol and feed markets, as shown in Figure I-3c, there is likewise no effect on market demand. The corn farmers, in this case, get no surplus from the subsidy despite what the government spends on it.

In general, corn farmers can benefit more than the amount the government spends on the subsidy only if the demand for ethanol is more elastic than the demand for feed. This is the empirically likely case, given that there are corn-free ways to make fuel that is essentially the same from the fuel consumer's perspective, but it is not as easy to switch to alternative animal feeds. Moreover, the supply of land for growing corn may be inelastic in the short run (but probably elastic in the long run).

How can we think about this intuitively? Think about price discrimination. Normally, we want to charge the low price to the people with elastic demand and the high price to people with the relatively inelastic demand. The ethanol subsidy looks like price discrimination precisely when the demand for ethanol is price elastic relative to feed because it pushes the ethanol price down relative to the feed price. Corn farmers can gain substantially in this scenario relative to spreading the same subsidy dollars across all corn sales.

We can also look at the equilibrium from the feed market perspective. Possible feed demand curves are already drawn in Figures I-1b, I-2b, and I-3b. The feed supply curve is a residual supply curve: the horizontal difference between the overall corn supply curve and the ethanol demand curve. The more elastic is ethanol demand, the more elastic the residual supply. In the perfectly elastic case introduced in Figure I-2, nothing is supplied to the feed market when prices are below the ethanol demand curve (all of the corn goes to ethanol)—and coincides with the overall supply curve at prices above that (no corn goes to ethanol). Figure I-4 therefore draws a supply curve that is horizontal at quantities in between the price axis and the overall supply curve.

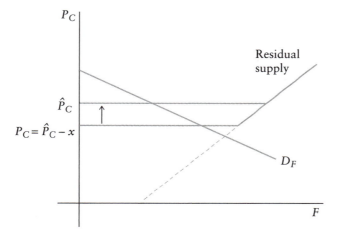

Figure 1-4: The supply of corn to feed usage is a residual supply curve. It is shifted up by the subsidy in the ethanol market. The case shown here corresponds to horizontal ethanol demand.

The ethanol subsidy x shifts up the residual supply curve by the amount x and raises the price that feed buyers pay for corn by x. The revenue that corn farmers gain in the feed market could easily exceed the revenue they gain in the subsidized market (ethanol) because (i) ethanol gets a minority of corn production and (ii) more important, ethanol demand is much more price sensitive than feed-corn demand.

The main idea here is that because we have a market, the subsidy on ethanol has an effect broader than its initial amount. The price of corn going into animal feed will also increase.

Price Theory Guides Measurement

In many labor, health, and other markets with large amounts of subsidies or taxes, there is a big difference between the price paid by buyers and the price received by sellers because one of the parties is paying a tax or receiving a subsidy. In these cases, price theory makes it obvious that the proper measurement of price depends on whether buyer or seller behavior is to be explained.

In our ethanol subsidy example, some buyers pay less than others. The use of the various prices for empirical analysis depends on the question at hand. For the purposes of predicting the amount of government revenue to subsidize corn sales, what matters is the quantity-weighted average subsidy in the market. That is the average of zero on feed corn and the subsidy rate on ethanol corn, weighted by the quantity of corn going to each use.

For the purposes of measuring the price impact, the quantity weights need to be adjusted for the price sensitivity of the buyers. In the neighborhood of no subsidy, the price impact formula is the product of three terms:[6]

$$\frac{dP_C}{dx} = \theta \; \frac{E}{C} \; \frac{P_C D'_E / E}{P_C D'_C / C} \; , \quad \theta = \frac{D'_C}{D'_C - S'}$$

where x is the subsidy rate, S' is the slope of the supply curve and θ is the usual incidence parameter indicating how each unit of a uniform subsidy would raise the price received by sellers. As a matter of algebra, we could further simplify the formula, but we keep the three terms separate in order to discuss their economic interpretation. The second term in the price impact formula is the quantity-weight term and recognizes that only a fraction (E/C) of the corn supplied goes to ethanol. The third term, with a price elasticity for both its numerator and denominator, adjusts for any difference between the ethanol demand elasticity and the overall demand elasticity. The third term ranges from zero when ethanol demand is completely inelastic (Figure I-3) to $C/E > 1$ when ethanol demand is infinitely elastic (Figure I-2); it would be one if both types of buyers were equally price elastic.[7]

In other words, the units sold to more-price-elastic buyers count more than the units sold to less-price-elastic buyers. In our example, with one type of buyer that is subsidized and the less price-sensitive type of buyer that is not, the price-sensitivity-adjusted weighted average subsidy exceeds the pure quantity-weighted average, which is why the corn farmers can gain more than the Treasury spends on the subsidy.

The analysis above refers to a subsidy rate that is small in comparison with the price. With larger subsidies we need to consider, for example, that the three terms in the formula vary with the level of the subsidy, which is essentially the price-index problem whose solutions are discussed in chapter 4.

EXAMPLE: ACQUIRED COMPARATIVE ADVANTAGE

With its emphasis on markets, price theory frequently highlights comparative advantage, which is about economic progress obtained through specialization and trade. The specialization made possible by markets helps explain where people live and work (Becker and Murphy 1992); why economies grow (Smith 1776/1904, Book I, Chapter I); why men are different from women (Becker 1985), but less so recently (Mulligan and Rubinstein 2008); and much more.

We examine the acquisition of comparative advantage in a simple market setup with two tasks, A and B. An individual has human capital for those tasks H_A and H_B. Whichever task is picked, a wage per unit of human capital is paid: w_A or w_B, as appropriate. This will mean total income for an individual from task A is $Y_A = w_A H_A$ and from task B is $Y_B = w_B H_B$. The maximum income that the individual can earn is

$$Y = \max\{w_A H_A, w_B H_B\},$$

which is obtained by picking task A if $w_A H_A > w_B H_B \Leftrightarrow \dfrac{w_A}{w_B} > \dfrac{H_B}{H_A}$, picking task B if $\dfrac{w_A}{w_B} < \dfrac{H_B}{H_A}$ and picking either task if the two ratios are equal. This is comparative advantage because the choice of task depends on the relative amounts of human capital held, not the absolute amount.

Figure I-5 illustrates the choice in the $[H_A, H_B]$ plane by drawing a solid task-indifference ray showing all of the configurations of human capital that someone could have and be indifferent toward the two tasks.

There is demand for tasks A and B, which in equilibrium has to match up with the available human capital and the aforementioned incentives for workers to choose one task rather than the other. This happens with wage adjustments. If there were a lot of demand for A, then Figure I-5's task-indifference ray has to be steep so that lots of workers choose task A and few choose B. In other words, w_A / w_B would be greater than 1.

Now, assume we have reached the equilibrium, so that w_A / w_B reflects market supply and demand. Then for any point on the line, every person directly below and directly left must be earning the same income. See the dashed lines in Figure I-5. This is because each person on the dashed line above the task-indifference ray has the same level of H_B and his or her H_A does not matter because it is not used. Each person on the dashed line below the task-indifference ray has the same level of H_A and H_B does not matter because he or she does not use it. Let's call the union of the two dashed lines an indifference curve for the worker.

Now, let's allow each agent to choose their human capital. For example, the agent is considering whether to be a good plumber versus being a good

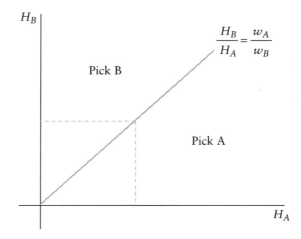

Figure I-5: Supply and demand will rotate the task-indifference ray until the right number of workers is in each task.

carpenter. The opportunity set for human capital could have an interesting shape, as depicted in Figure I-6. Consider the point associated with the maximum level of H_B. As it is depicted, this person will have some positive level of H_A. This reflects an underlying story that tasks A and B require some of the same abilities. Thus, if I choose to be a good plumber, that doesn't mean that I end up with zero human capital as a carpenter.

Note further that in this graph, the economically relevant region of the opportunity set lies between the two points, and we can erase the parts of the curve close to the axes because no one would choose a human capital pairing left of the top point or below the right point. On the erased regions, the agent could be better at both tasks!

Now let's put the opportunity set together with the worker's indifference curves, as in Figure I-7. We can even have everyone identical in the sense that they all have the same opportunity curve to choose from. Nevertheless, specialization is optimal behavior. Being equally good at tasks A and B is worse than being very good at just one task because you have acquired a lot of human capital that you do not use.

We started this picture by indicating the types of workers (that is, configurations of human capital) who are indifferent between the two tasks. But now we have shown that people will not choose to be those types of workers. Because human capital is acquired, indifference toward the two tasks does not occur in equilibrium.[8]

The equilibrium requires that both tasks are performed, so some people specialize in A and others in B. People who are identical, in the

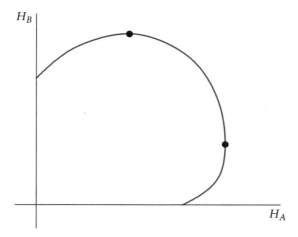

Figure I-6: The opportunity set for selecting human capital. The agent with maximum human capital for task *A* still has positive human capital for task *B*.

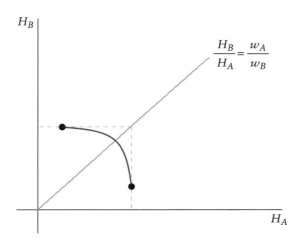

Figure I-7: Specialization. Agents maximize their human capital at task *A* or task *B*.

sense of having the same opportunities open to them, end up being different.

You might say that it is a coin flip as to exactly who goes toward task *A* and who toward task *B*, and we would agree if people were precisely identical. But in reality, people have somewhat different opportunities open to them: in Figures I-6 and I-7, that means somewhat different opportunity curves. Some of the opportunity curves may be relatively steep and others relatively flat. Then just a small difference among people in the slope of the curve will decide who specializes in what. Specialization in the marketplace turns small differences into large ones.

OUTLINE OF THE COURSE

Three economic themes are repeatedly encountered when human behaviors are viewed through the lens of economic theory: substitution effects, market equilibrium, and durable goods. Each of these is a part of the course presenting the classic model and then going through some important applications such as price indices, learning by doing, and house prices.

Part I, on prices and substitution effects, is written from the perspective of consumer theory. We see little need to explicitly treat firms here, merely for the sake of repetition. The theory of substitution effects is the foundation of price and quantity indices (chapter 4), which are among the most widely used tools for economic measurement. Chapter 5 looks at a bit of "behavioral economics" from the perspective of the Marshallian demand curve. The distinction between short- and long-run demand, examined in chapter 6, has a number of immediate and nontrivial applications such as habits and addictions.

Once we have consumers, the purpose of bringing in firms is to have markets (Part II), which are the primary emphasis of the course. Here we begin with Adam Smith's (1776/1904) compensating differences, as further developed by Sherwin Rosen (1986) in his publications and teaching price theory at Chicago. Without saying much yet about production, this allows us to obtain results for urban economics and the accumulation of human capital.

One of the lessons of compensating differences is to be wary of purported "free lunches." The learning-by-doing application is of significant intrinsic interest but was also one of Becker's and Rosen's favorite demonstrations of a consequence of market competition, which reappears in a great many applications, ranging from health insurance to industrial organization to taxation.

Firms are carefully examined toward the end of Part II. This completes the foundation of the "industry model" (aka supply and demand), thereby opening up a huge range of applications. One application with particularly surprising results is the consequence of prohibiting trade in specific goods such as illegal narcotics, which is the subject of chapter 12. Exclusive dealing, quantity discounts, and other pricing practices are also readily examined once we have consumers and firms together, as we show in chapter 13. The final chapter of Part II extends the industry model to more than two production factors, which is helpful for examining durable goods (as in Part III).

Part III looks at changes over time. It begins by defining durable goods and extending the industry model to include both a capital-rental market and a capital-purchase market (chapters 15 and 16). This brings us pretty close to the adjustment-cost model of investment and the neoclassical growth model (chapter 17). These are usually considered "macroeconomics" topics but, as factor supply and demand repeat over time, the two models should not be omitted from price theory. Most important, price theory treats durable goods because durability is an important feature in many practical questions.

The final three chapters look at important applications of the durable goods models—such as capital-income tax incidence, the determination of labor's share of national income, and investments in health.

Part I

Prices and Substitution Effects

Chapter 1

Utility Maximization and Demand

UTILITY MAXIMIZATION

We develop the analysis of substitution effects with consumer theory, and leave producer theory until the section on market equilibrium. Rather than taking the axiomatic approach that is typically presented in microeconomics textbooks, we follow the more practical approach found in most applied work. That is, we start with a utility function defined over a set of goods $X_1, \ldots, X_N$, denoted $U(X_1, \ldots, X_N)$. We assume that each of these goods is being purchased in a market. Later, we think about goods produced at home, purchased in another way, or goods a person is endowed with. We further assume you are purchasing these goods at prices $P_1, \ldots, P_N$. Finally, we assume the agent has income M. Note that we need units to be consistent. If income M is income per week, then consumption X is consumption per week. These are both flows.

We will usually think about consumers having some budget constraint

$$\sum_{i=1}^{N} X_i P_i \le M.$$

That is, the agent cannot spend more than his or her income. Later, in other models, we will think about time constraints. Now, we consider a problem of maximizing utility subject to a budget constraint

$$\max_{X_1, \ldots, X_N} U(X_1, \ldots, X_N)$$

$$s.t. \ \sum_{i=1}^{N} X_i P_i \le M.$$

One of the toughest parts of the theory is that we don't have much ex ante information about what utility looks like. This will be much different than production, where we see the inputs and then how much output those inputs make. We therefore want to minimize the role of utility in the analysis.

We set up the Lagrangian

$$L = U(X_1, \ldots, X_N) + \lambda \left[M - \sum_{i=1}^{N} X_i P_i \right].$$

The problem is solved by taking the derivative of the Lagrangian with respect to the X's and λ. This procedure gives the first-order conditions

$$\frac{\partial U}{\partial X_1} - \lambda P_1 = 0$$

$$\ldots$$

$$\frac{\partial U}{\partial X_N} - \lambda P_N = 0$$

$$\sum_{i=1}^{N} X_i P_i = M.$$

It follows that

$$\frac{\dfrac{\partial U}{\partial X_i}}{\dfrac{\partial U}{\partial X_j}} = \frac{P_i}{P_j}.$$

What is the right-hand side? This tells us that to get one unit of good i, we *must give up* P_i / P_j units of good j. More generally, we call P_i / P_j the cost of good i (in units of j). What is the left-hand side? This tells us the amount we're *willing to give up* of good j to get a unit of good i. More generally, we call the left-hand side the value of good i (in units of j). So, all this expression really says is that to be at an optimal point, it must be the case that the cost of consuming more of good i equals your value for consuming more of good i. This expression is commonly explained as "marginal benefit must equal marginal cost."

See Figure 1-1 for a graphical depiction of the optimum in the two-good case. The tangency point is the consumption choice that equates the utility cost of shifting between the two goods with the budget cost.

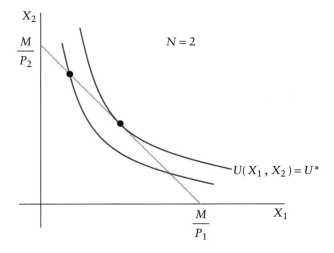

Figure 1-1: The point given by the intersection of the budget constraint with the indifference curve labeled by U^* is the optimum. The second point depicted is not an optimum because the agent is still getting more marginal utility per dollar for units of good X_1 than X_2.

The second point shows the indifference curve crossing the budget constraint, which means that utility can be improved by shifting to good 1 without violating the budget.

The reason the proportional relationship of prices and marginal utility is important is the fact that we can measure prices directly. We can infer something about preferences from observed behavior. The assumption that people are maximizing gives us a lot of insight into preferences, which we cannot directly measure. We do not have a theory of preferences that tells us what people like before they choose certain products, so instead we use an empirical theory of preferences: people like what they choose.

The marginal benefit and cost equation can also be written as

$$\frac{\frac{\partial U}{\partial X_i}}{P_i} = \frac{\frac{\partial U}{\partial X_j}}{P_j} = \lambda.$$

This version says that, at the optimum, the utility per dollar on the margin must be equal for all goods. This makes sense intuitively. Suppose the marginal utility per dollar was higher for good i than good j. Then you'd be better off by giving up some of good j and consuming more of good i. This expression also tells us that λ is the marginal utility per dollar. Importantly, we don't really need to worry about how that dollar is spent. We can talk about value independent of what a person actually does. For example, suppose dentists are considering adding fluoride to the water supply to reduce cavities. It may be that once fluoride is in the water, people stop brushing their teeth and they get as many or more cavities as before. Dentists might say that the policy backfired and people didn't benefit. The economist would say: sure, people benefited; now they don't have to brush their teeth. It doesn't matter how they choose to take the benefit of fluoride in the water supply—by getting fewer cavities or by not having to brush their teeth.

We can say something else about λ. Since our first-order conditions give

$$\frac{\partial U}{\partial X_i} = \lambda P_i,$$

for every i, we have that (at the optimum) marginal utilities are proportional to prices, and that proportion is λ. So if we can measure the prices people face, we can indirectly measure their marginal utilities because the marginal utilities are proportional. Let's say the price of good 10 is five times the price of good 6, so that $P_{10} = 5P_6$. This tells us that a unit of good 10 is worth five times as much as good 6 on the margin. This is true for everyone in the market consuming goods 6 and 10. It's not just a "market value."

Now let's consider two states of the world. In state 1, people consume goods $X_1^*, \ldots, X_N^*$ at prices $P_1, \ldots, P_N$ and have income M. Now we perturb everything slightly. That is, change consumption by $dX_1, \ldots, dX_N$, prices by $dP_1, \ldots, dP_N$, and income by dM. Then how does utility change? That is, what is dU? We know $dU = dU(X_1, \ldots, X_N) = \sum_{i=1}^{N} \frac{\partial U}{\partial X_i} dX_i = \lambda \sum_{i=1}^{N} P_i dX_i$. The key is that $P_i dX_i$ is observable because we know the prices and the changes in consumption. Then we know $\frac{dU}{\lambda} = \sum_{i=1}^{N} P_i dX_i$. But the left-hand side is just the dollar value of how much better off the consumer is. What's at the core of this analysis? Because people are optimizing, the marginal utilities of goods are proportional to their prices.

Similarly, we can consider how utility changes over time. Consider time-varying consumption $X_1(t), \ldots, X_N(t)$. Then $\frac{dU}{dt} = \sum_{i=1}^{N} \frac{\partial U}{\partial X_i} \frac{dX_i}{dt} = \sum_{i=1}^{N} \lambda P_i \frac{dX_i}{dt}$, where in the last step we once again use the fact that marginal utilities are proportional to their prices. Thus, the dollar value of the change in utility is

$$\frac{\frac{dU}{dt}}{\lambda} = \sum_{i=1}^{N} P_i \frac{dX_i}{dt}.$$

In terms of Figure 1-1's tangency illustration, we can think of this formula as measuring the utility change from one optimum to another according to the extra income required to achieve the new utility level. We return to this measurement approach when we look at price and quantity indices in chapter 4.

THE THEORY OF DEMAND

The solutions to the first-order conditions, $X_1^* = X_1^M(P_1, \ldots, P_N, M), \ldots,$ $X_N^* = X_N^M(P_1, \ldots, P_N, M), \lambda^* = \lambda(P_1, \ldots, P_N, M)$, are known as demand equations. They're a particular type of equation system called the Marshallian demand equations, named after Alfred Marshall.

Applied work often begins with demand equations rather than explicitly deriving them from utility functions. But not every set of equations relating prices to the quantities purchased by an individual are consistent with utility maximization; our purpose here is to show what is special about demand equations.

Note that the Marshallian demand function allows the agent to alter consumption of all goods in response to the price change; for example, we can consider the effect of changing P_j on good 1; that is, $\partial X_1^M / \partial P_j$. In taking these to the data, we might get differences between short-run and long-run responses because it takes time for people to adjust in reality (see chapter 6).

Utility maximization places some restrictions on these Marshallian demand equations. These restrictions are most easily expressed in terms of demand elasticities. First, ϵ_{ii}, the elasticity of demand for good i with respect to the price of good i is given by

$$\epsilon_i = \epsilon_{ii} = \frac{\% \Delta X_i}{\% \Delta P_i} = \frac{P_i}{X_i} \frac{\partial X_i^M}{\partial P_i} = \frac{\partial X_i^M}{X_i} \bigg/ \frac{\partial P_i}{P_i}.$$

This is also called own-price elasticity and often shortened from ϵ_{ii} to ϵ_i. It gives us the percent change in demand for good i for each 1% increase in the price of good i. We also have the cross-price elasticity,

$$\epsilon_{ij} = \frac{P_j}{X_i} \frac{\partial X_i^M}{\partial P_j},$$

which gives the percent increase in demand for good i for each 1% increase in the price of good j. Typically, if $\epsilon_{ij} > 0$—demand for good i increases when the price for good j increases—we say that goods i and j are substitutes. If $\epsilon_{ij} < 0$, goods i and j are complements.

Lastly, the income elasticity of demand for good i is

$$\eta_i = \frac{M}{X_i} \frac{\partial X_i^M}{\partial M}.$$

This gives us the percent change in demand for good i in response to each 1% change in income. For $\eta_i > 0$, we say i is normal. It natural to think about an income elasticity of 1, because it corresponds to a 10% increase in income resulting in scaling up consumption of everything by 10%. This is the threshold between luxury and necessity goods: for $\eta_i < 1$, i is a necessity; for $\eta_i > 1$, i is a luxury. This intuition also suggests that the average income elasticity is 1:

$$\sum_{i=1}^{N} s_i \eta_i = 1,$$

where s_i is the share of good i in total spending (aka, good i's "budget share"). Note that the shares and elasticities are evaluated at a particular set of prices and a particular income. The fact that the share-weighted income elasticity of demand is 1 is sometimes known as "Engel aggregation."

We can see mathematically that the average income elasticity is 1 by (a) differentiating the budget constraint with respect to income, (b) rewriting each income-effect term with the corresponding income elasticity, (c) rewriting each price term with the corresponding spending share, and (d) canceling like terms in numerators and denominators:

$$1 = \sum_i \frac{\partial X_i^M}{\partial M} P_i = \sum_i \left(\frac{X_i^M \eta_i}{M} \right) \left(s_i \frac{M}{X_i^M} \right) = \sum_i \eta_i s_i.$$

We have two more constraints on elasticities, namely

$$\sum_i s_i \epsilon_{ij} = -s_j$$

and

$$\sum_j \epsilon_{ij} + \eta_i = 0.$$

The first, known as the "adding up" constraint (for the Marshallian demand system), is a sum over goods.[1] It can be proved by differentiating the budget constraint with respect to the jth price and, as before, rewriting demand slopes in terms of elasticities and prices and quantities in terms of shares:

$$\sum_i \frac{\partial X_i^M}{\partial P_j} P_i + X_j^M = 0 \Leftrightarrow \sum_i \left(\frac{X_i^M}{P_j} \epsilon_{ij} \right) \left(s_i \frac{M}{X_i^M} \right) = -\frac{M}{P_j} s_j$$
$$\Leftrightarrow \sum_i s_i \epsilon_{ij} = -s_j.$$

Because quantities are not negative, this constraint is another way of expressing a law of demand.[2] Increasing the price of good j may increase the Marshallian demand for good j (specifically, when good j is a Giffen good) but then it must, weighted by prices, reduce the demand for other goods even more.

The second elasticity constraint, known as "homogeneity" (for a Marshallian demand function), is a sum over prices merely saying that increasing income and all prices by the same proportion has no effect on choices, expressed in terms of elasticities. Take, for example, a strictly positive demand for good i in response to a $\mu \neq 0$ proportion change in prices and income:

$$0 = \sum_j \frac{\partial X_i^M}{\partial P_j} P_j \mu + \frac{\partial X_i^M}{\partial M} M\mu = \mu X_i^M \left(\sum_j \epsilon_{ij} + \eta_i \right).$$

What's an example of an inferior good? Classic examples include potatoes and Kraft Mac and Cheese. But is food inferior? In general, it is not. As income increases, people buy "more" food. But what do we mean by "more?" We mean $FOOD = \sum_{i \in FOOD} P_i X_i$, and then

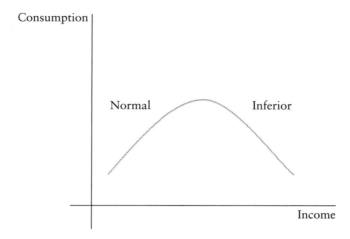

Figure 1-2: Goods can be normal over certain income ranges and inferior over others.

$dFOOD = \sum_{i \in FOOD} P_i dX_i$. We're not measuring more food in calories, pounds, etc.—we're measuring it by weighting the consumption by prices. This is useful because P_i gives us insight into the agent's value for X_i. We could easily have total calories or total pounds of food consumption decreasing while our price-based measure of food increases.

Note also that whether goods are normal or inferior is highly dependent on the level of income. In many areas of the world, Walmart expects demand for its products to increase when income rises. In the United States, on the other hand, consumers facing higher income might be more likely to stop shopping at Walmart and shop elsewhere instead. Figure 1-2 illustrates with an Engel curve—a graph of income versus quantity demanded—a good that is normal at lower incomes and inferior at relatively high incomes. If the good were shopping at Walmart, then the peak demand would happen at the income at which people tend to shift their shopping to "fancier" stores.

The analysis so far has taken U as just a black box. But take cars and gasoline, for example. Are they represented as complements in the utility function because of psychology? No, not at all. Cars and gasoline are complements because they are both needed for transportation. It has to do with technology much more than preferences.

This theory we've developed is not a heuristic or a tautology. For instance, suppose we know a person chose to consume $X_1, \ldots, X_N$ at prices $P_1, \ldots, P_N$ and $\sum P_i X_i = M$. Then we know that person would prefer

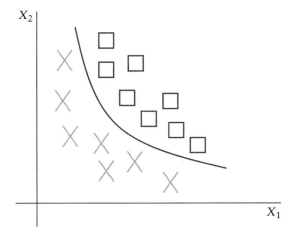

Figure I-3: Boxes denote choices preferred over bundles on the indifference curve, and crosses denote choices not preferred over bundles on the indifference curve.

that point to $\hat{X}_1, \ldots, \hat{X}_N$, where $\sum P_i \hat{X}_i < M$ because the bundle of $\hat{X}$'s was originally affordable but was not chosen. Further, by looking at real choices, we can back out indifference curves, as in Figure 1-3.

It is important that the choices are "real," however. Do not ask a man who is an alcoholic how much he drinks because he will likely understate his consumption. Look at choices people actually make in the marketplace. Those are the choices depicted in Figure 1-3.

Chapter 2

Cost Minimization and Demand

Chapter 1 derived the Marshallian demand curves for a consumer with income M choosing among N goods, which are repeated below for the reader's convenience.

$$X_1 = X_1^M(P_1, \ldots, P_N, M)$$
$$\ldots$$
$$X_N = X_N^M(P_1, \ldots, P_N, M).$$

Note that there are no quantities in the Marshallian demand functions. Any price change analyzed with this system involves changes in the quantities consumed of every good. A change in the price of toys not only affects the number of toys purchased for each child, but the number of children! In turning to the data, it is important to ask what is really varying: it may be that some of the quantities are held fixed, in which case there is a quantities-constant system (more on this later) to use for the analysis.

THE COST FUNCTION

Consider a cost-minimization problem, the dual of the utility maximization problem. This problem is

$$\min_{X_1, \ldots, X_N} \sum_{i=1}^{N} X_i P_i$$
$$s.t. \ U(X_1, \ldots, X_N) = \bar{U}.$$

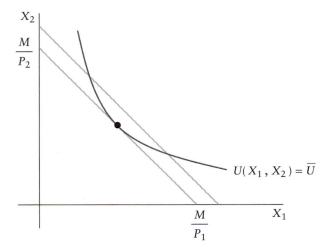

Figure 2-1: The primal and dual approaches to have the same consumer problem first-order conditions. Finding the highest indifference curve for a given budget constraint is closely related to finding the lowest budget constraint for a given indifference curve.

Instead of looking at the choices that maximize the level of utility that can be obtained with a given income, we now look at the choices that minimize the expenditure needed to achieve a given level of utility. For the maximization problem examined in chapter 1, the budget constraint was fixed and we looked for the consumption choices that attain the highest indifference curve reachable with that budget. In the minimization problem, we fix the indifference curve and find the consumption choices that attain that utility using the least possible income, as in Figure 2-1. These procedures land us at the same optimum point.

The Lagrangian for this problem is

$$L = \sum_{i=1}^{N} X_i P_i + \mu[\bar{U} - U(X_1, \ldots, X_N)].$$

By taking derivatives and setting them equal to 0, we get the first-order conditions

$$P_1 = \mu \frac{\partial U}{\partial X_1}$$

$$\ldots$$

$$P_N = \mu \frac{\partial U}{\partial X_N}$$

$$\bar{U} = U(X_1, \ldots, X_N).$$

Except for the last one, these are the same conditions as before; we just have $\mu = 1/\lambda$. The first-order conditions from cost minimization give the Hicksian demand functions

$$X_1^* = X_1^H(P_1, \ldots, P_N, \bar{U})$$
$$\cdots$$
$$X_N^* = X_N^H(P_1, \ldots, P_N, \bar{U}).$$

These choices resulting from the cost minimization problem are no different than the choices we got from the maximization problem. That is, $X_1^* = X_1^H(P_1, \ldots, P_N, \bar{U}) = X_1^M(P_1, \ldots, P_N, M)$. The only difference is that now we're indexing by the level of utility achieved rather than the level of income required.

As economists, we want to use this theory to make predictions. One result from the cost minimization problem that is particularly useful is the cost function—the minimum cost needed to achieve a level of utility $\bar{U}$ given prices $P_1, \ldots, P_N$, is

$$C(P_1, \ldots, P_N, \bar{U}) = \min \sum_{i=1}^N X_i P_i \ \ s.\,t. \ U(X_1, \ldots, X_N) = \bar{U}.$$

The cost function has a few properties that make it a useful tool for solving problems in economics. It is homogeneous of degree 1 in prices. If all prices double, the cost doubles. It is nondecreasing in prices; that is, $\frac{\partial C}{\partial P_i} \geq 0$. Furthermore, the partial derivative of the cost function is the Hicksian demand curve, $\frac{\partial C}{\partial P_i} = X_i^H(P_1, \ldots, P_N, \bar{U})$. The cost function is also concave in prices.[1] Finally, cost is increasing in utility; that is, $\frac{\partial C}{\partial U} > 0$. To summarize:

Properties of the Cost Function

1. C is homogeneous of degree 1 in prices
2. $\frac{\partial C}{\partial P_i} \geq 0 \ \forall i$
3. $\frac{\partial C}{\partial P_i} = X_i^H(P_1, \ldots, P_N, \bar{U})$
4. C is concave in prices
5. $\frac{\partial C}{\partial U} > 0.$

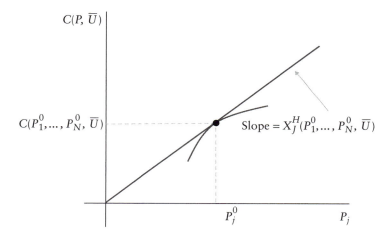

Figure 2-2: Graphical proof of the concavity and derivative properties of the cost function. The line depicts the "monkey" solution of just buying the same bundle as the price of good j differs from P_j^0, and the curve depicts the cost function.

We prove properties (3) and (4) intuitively.

Let's consider a unidimensional case, as in Figure 2-2. In our experiment, we move the price of good j from some initial price P_j^0, while all other prices $P_1^0, \ldots, P_{j-1}^0, P_{j+1}^0, \ldots, P_N^0$ are held constant. Consider the point $(P_j^0, C(P_1^0, \ldots, P_N^0, \bar{U}))$, and suppose we consider the "monkey" solution where, as the price of good j changes, we do not change our consumption bundle at all. Then we'd move straight along a line that is linear in P_j. If we're consuming 10 units of good j, then for every dollar increase in P_j, we spend 10 more dollars. The slope of that line equals $X_j^H(P_1^0, \ldots, P_N^0, \bar{U})$. It's just the demand bundle, so it's also equal to the Marshallian quantity at those prices and the associated level of income. But we know only the point $(P_j^0, C(P_1^0, \ldots, P_N^0, \bar{U}))$ is optimal on the line with slope $X_j^H(P_1^0, \ldots, P_N^0, \bar{U})$ because the optimal bundle adjusts as P_j changes. Other optimal points are therefore below the line associated with the "monkey" strategy we've considered—a point on the line gives a feasible cost for achieving $\bar{U}$ and the minimum cost must be below or equal to any feasible cost. But how concave is the line associated with the optimal cost? It is more concave the more room for adjustment there is.

This also connects the cost function to the elasticity of demand. We have just considered, intuitively, why $\dfrac{\partial C}{\partial P_j} = X_j^H(P_1, \ldots, P_N, \bar{U})$, but we also have that $\dfrac{\partial^2 C}{\partial P_j^2} = \dfrac{\partial X_j^H(P_1, \ldots, P_N, \bar{U})}{\partial P_j}$. In other words, the function can only be as concave as the demand function is responsive to a change in price.

HICKS' GENERALIZED LAW OF DEMAND

There's another way to think about the concavity of the cost function. Consider two price-utility vectors with the same utility level, $(P_1^0, \ldots, P_N^0, \bar{U})$ and $(P_1^1, \ldots, P_N^1, \bar{U})$. Assume the optimal quantities associated with these vectors are $X_1^0, \ldots, X_N^0$ and $X_1^1, \ldots, X_N^1$. The fact that these bundles are cost minimizing tells us that

$$\sum_{i=1}^{N} X_i^1 P_i^0 \geq \sum_{i=1}^{N} X_i^0 P_i^0.$$

Similarly, $\sum_{i=1}^{N} X_i^0 P_i^1 \geq \sum_{i=1}^{N} X_i^1 P_i^1$ by the same logic. We can add these two inequalities to get

$$\sum_{i=1}^{N} X_i^1 P_i^0 + \sum_{i=1}^{N} X_i^0 P_i^1 \geq \sum_{i=1}^{N} X_i^0 P_i^0 + \sum_{i=1}^{N} X_i^1 P_i^1.$$

Simplifying this condition, we get

$$\sum_{i=1}^{N} (X_i^1 - X_i^0)(P_i^1 - P_i^0) \leq 0.$$

This is a generalized version of the law of demand, also credited to Sir John Hicks.[2] It's more general than $\dfrac{\partial X_i^H}{\partial P_i} \leq 0$ because the generalized version has all of the prices changing at the same time, the price changes do not have to be infinitesimal, and the Hicksian demand curves need not have derivatives at the points of interest. On average, goods whose prices go up are consumed less (and the opposite is also true). Equivalently, the cross-good correlation between price changes and demand changes cannot be positive. So the law of demand comes right out of cost minimization. You can't say "it got cheaper and my effective income hasn't changed, but I'm still going to buy less." This would be inconsistent with the idea of cost minimization.

The generalized law of demand says that the Hicksian demand system is concave. The law of demand for individual Hicksian demand curves—that is, $\dfrac{\partial X_i^H}{\partial P_i} \leq 0$—says that, good-by-good, quantities are non-increasing in their prices.

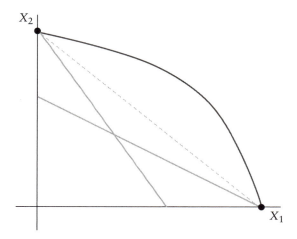

Figure 2-3: Indifference curves concave to the origin yield solutions no different than a linear indifference curve. Consumers pick optimal points at the corners depending on how steep the budget constraint is.

The cost function also shows that the cross-price effects on Hicksian demand are symmetric, $\dfrac{\partial X_i^H}{\partial P_j} = \dfrac{\partial X_j^H}{\partial P_i}$, since both are equivalent to $\dfrac{\partial^2 C}{\partial P_i \partial P_j}$. This is *not* a statement that the elasticities are equal; in general, the elasticities differ systematically from equality.

RELATIONSHIPS BETWEEN INDIFFERENCE CURVES AND THE DEMAND SYSTEM

Cost and demand functions are related to the indifference curves. As the indifference curves become more curved, demand becomes more inelastic but the cost function becomes straighter. So the cost function is closer to linear when the indifference curves have significant curvature. The opposite is also true. Suppose we have the linear-in-prices cost function $C(P_1, P_2, \bar{U}) = a(\bar{U})P_1 + b(\bar{U})P_2$. What do the indifference curves look like? They're a right angle at the point (a, b). What if $C(P_1, P_2, \bar{U}) = \min(a(\bar{U})P_1, b(\bar{U})P_2)$? Then the indifference curves are straight lines.

What about someone whose indifference curves are bowed outward from the origin? See Figure 2-3. If budget lines are steep, the agent picks only the y-axis object; if they're flat, the consumer picks only the x-axis object. Empirically, these agents look just like the consumers with linear

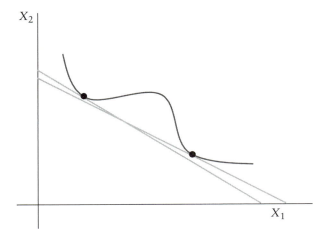

Figure 2-4: A region of nonconvex preferences creates a price threshold around which the consumer makes a large change in consumption in response to a small change in relative prices.

indifference curves (such as the dashed line shown in the figure) because they only pick the endpoints.

Consider an indifference curve that has a more typical shape but has a region concave to the origin like the previous case. See Figure 2-4. Then, for certain very small price movements, the consumer makes large changes in his or her consumption bundle.

PROPERTIES OF HICKSIAN DEMAND FUNCTIONS

The fact that the cost function is homogenous of degree 1 in prices means that each Hicksian demand function is homogenous of degree 0 in prices. That is, $X_i^H(\alpha P_1, \ldots, \alpha P_N, \bar{U}) = X_i^H(P_1, \ldots, P_N, \bar{U})$. Because these functions are equivalent for arbitrary α, their derivatives are also equal, which means $\sum_{j=1}^{N} \frac{\partial X_i^H}{\partial P_j} P_j = 0$. Dividing both sides by X_j and denoting the Hicksian cross-price elasticity of good i demand with respect to P_j as ϵ_{ij}^H, we get

$$\sum_{j=1}^{N} \epsilon_{ij}^H = 0.$$

This says that the sum of all of good i's cross-price elasticities is 0. But symmetry also gives us some intuition into elasticities. We can multiply and divide by 1 a few times to get from $\dfrac{\partial X_i^H}{\partial P_j} = \dfrac{\partial X_j^H}{\partial P_i}$ to $\dfrac{X_i P_i}{M} \dfrac{P_j}{X_i} \dfrac{\partial X_i^H}{\partial P_j}$ $= \dfrac{\partial X_j^H}{\partial P_i} \dfrac{P_i}{X_j} \dfrac{P_j X_j}{M}$. Using the income shares for each good (recall that $s_i = \dfrac{X_i P_i}{M}$) we get $s_i \epsilon_{ij}^H = s_j \epsilon_{ji}^H$. Then the ratio of the cross-elasticities is equivalent to the ratio of the shares. In general, this says that the ϵ_{ij}^H is not the same as ϵ_{ji}^H. But it says more than that. Suppose $s_j > s_i$. Then it's clear that the big good j matters more than the small good i—the elasticity of demand for good i with respect to the price of good j is larger than the elasticity of demand for good j with respect to the price of good i.

Now we can think about utility again. Note, by construction, we have

$$ U(X_1^H(P_1, \ldots, P_N, \bar{U}), \ldots, X_N^H(P_1, \ldots, P_N, \bar{U})) = \bar{U}. $$

But this holds for any P, so we can differentiate with respect to P_i to get

$$ \sum_{j=1}^{N} \frac{\partial U}{\partial X_j^H} \frac{\partial X_j^H}{\partial P_i} = 0 \underset{\text{F.O.C.}}{\Rightarrow} \sum_{j=1}^{N} P_j \frac{\partial X_j^H}{\partial P_i} = 0. $$

That is, we can once again see that marginal utilities are proportional to prices. But this gives us a corollary in elasticity form, namely that

$$ \sum_{j=1}^{N} s_j \epsilon_{ji}^H = 0. $$

This sum over goods is called "adding up" (for the Hicksian demand system). It adds the effects of a single price change (P_i) on all demands ($j = 1 \ldots N$). Homogeneity, on the other hand, was about a single demand equation: it adds all of the price effects in that equation. If we change all prices by the same percentage, we get no change in consumption. Adding up says that, looking across all equations, whenever a single price changes, all the goods have to change in such a way that, weighted by their shares, the total change is 0. So, three restrictions emerge: homogeneity, symmetry, and adding up. Note (convince yourself of this!) that symmetry and homogeneity imply adding up, and adding up and symmetry imply homogeneity. In the two-good case, adding up and homogeneity imply symmetry, but this does not hold in general.

Chapter 3 links the Hicksian and Marshallian systems using the Slutsky equation.

Chapter 3

Relating the Marshallian and Hicksian Systems

Chapters 1 and 2 covered two different demand systems, the Marshallian approach—maximizing utility subject to a budget constraint—and the Hicksian approach—minimizing cost subject to a utility constraint. Because they're two ways of looking at the same problem, we are not limited to using the tool that corresponds to the problem the consumer is solving. That is, if we know a consumer is solving the utility maximization problem subject to a budget constraint, then, as analysts, we can still use the tools of the Hicksian approach. For many problems, the Hicksian approach will prove very useful. Throughout the book we use both methods, picking the one better suited to solving the problem at hand.

THE SLUTSKY EQUATION

It's convenient to know how to go back and forth between the Hicksian and Marshallian systems. See Figure 3-1. Suppose a cost-minimizing consumer picks the optimal point (X_1^*, X_2^*). Then $X_1^* = X_1^H(P_1, P_2, \bar{U})$. But now suppose we give the consumer a level of income equal to $C(P_1, P_2, \bar{U})$ and he or she solves the utility maximization problem rather than the cost minimization problem. The consumer will choose the same optimal point (X_1^*, X_2^*). That is, $X_1^* = X_1^H(P_1, P_2, \bar{U}) = X_1^M(P_1, P_2, C(P_1, P_2, \bar{U}))$. We'll call this the Slutsky correspondence. This is one level above the Slutsky equation, which is about derivatives.

But note that the Slutsky correspondence is *not* an equilibrium condition determining a price or utility level. Rather, the Hicksian and

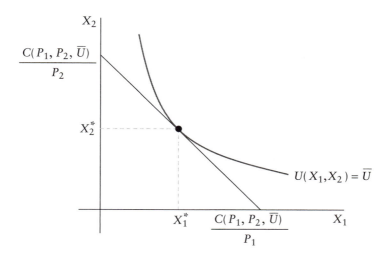

Figure 3-1: A graphical depiction of the idea that $X_1{}^* = X_1{}^H (P_1, P_2, \bar{U}) = X_1{}^M (P_1, P_2, C(P_1, P_2, \bar{U}))$.

Marshallian functions are the same when the latter has income replaced by the cost function: the equation holds for every price and every utility level.

More generally, the Slutsky correspondence for good i is $X_i^H(P_1, \ldots, P_N, \bar{U}) = X_i^M(P_1, \ldots, P_N, C(P_1, \ldots, P_N, \bar{U}))$. Then we can take the derivative with respect to the jth price and use the fact that each price derivative of the cost function is equal to the corresponding quantity demanded $\left(\dfrac{\partial C}{\partial P_j} = X_j \right)$, to get

$$\frac{\partial X_i^H}{\partial P_j} = \frac{\partial X_i^M}{\partial P_j} + \frac{\partial X_i^M}{\partial M} X_j.$$

This is the Slutsky equation. It lets us go back and forth between the derivatives of the Hicksian and Marshallian systems, and it tells us how income-constant changes in price are related to utility-constant changes in price. The difference between them comes from the rightmost term, called the income effect. Let's rearrange this expression to be $\dfrac{\partial X_i^M}{\partial P_j} = \dfrac{\partial X_i^H}{\partial P_j} - \dfrac{\partial X_i^M}{\partial M} X_j$. The left-hand side is the Marshallian effect, the first term on the right-hand side is the substitution or Hicksian effect, and the last term is the income effect.

What does the Slutsky equation say intuitively? Suppose P_j increases by a dollar. There will be a response holding utility constant, which is the substitution effect. Even though we call this the "substitution effect," it could be either a substitution relationship or a complementarity relationship. But there is also an effect coming from the fact that the change in price has changed our income. If we were consuming 10 of good j, then we have, in some sense, 10 fewer dollars of income. That is, to buy what I bought before would cost 10 more dollars after the price hike. The income effect tells us how responsive good i is to income. That is, if my income changes by 10 dollars, how much does my demand for good i change?

The Marshallian and Hicksian price effects differ by the income-effect term. The income-effect term tends to be small for most goods because most goods are a small share of the overall household budget. If I make \$100,000/year but only become \$10 poorer as a result of the good j's price change, the income effect on my demand for good i will not be large.

To see all of this more clearly, let's translate the Slutsky equation into elasticity form. Multiply by $\dfrac{P_j}{X_i}$ on both sides and by $\dfrac{M}{M}$ on the term designating the income effect to get

$$\epsilon_{ij}^M = \epsilon_{ij}^H - s_j \eta_i.$$

This is the elasticity version of the Slutsky equation. It says that the Marshallian percentage response of a percent increase in the price of good j on good i is the Hicksian response to the price change, minus the income share of the good whose price has changed times the income elasticity of the good we're looking at. Thus, the size of the income effect depends on the share of good j and how responsive to income good i is.

This is where the law of demand can run into some problems. Let's consider the own-price case, where $j = i$. Then

$$\epsilon_{ii}^M = \epsilon_{ii}^H - s_i \eta_i.$$

Theory tells us that $\epsilon_{ii}^H \leq 0$. Then, if the good is normal, $\eta_i > 0$, and the share is positive, we're guaranteed that $\epsilon_{ii}^M < 0$ so that the law of demand holds. Moreover, for normal goods, the Hicksian demand curve is downward sloping, but the Marshallian demand curve is even more responsive to price because the income effect reinforces the substitution effect.

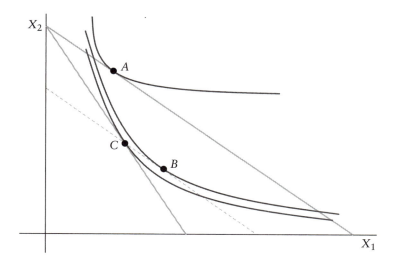

Figure 3-2: After an increase in P_1, consumption moves from A to C. The move from A to B denotes the income effect part of the change in consumption. As income falls, much more of X_1 is consumed: good 1 is inferior. The move from B to C is the substitution effect: consuming less of the good that got more expensive. The figure shows the Giffen case because an increase in P_1 has an income effect on good 1 that more than offsets its substitution effect.

The negativity of ϵ_{ii}^M could be violated, however, if the good has a large share and is inferior, so that $\eta_i < 0$. This case is called a *Giffen good*, where typically we have a weak substitution effect and a large-share inferior good. These are very difficult to find because most high-share goods are normal. See Figure 3-2 for a depiction of the Giffen good case.

ADDING UP AND SYMMETRY FOR THE MARSHALLIAN SYSTEM

The Slutsky equation also allows us to move between the conditions on Marshallian and Hicksian demand. In the Marshallian system, we have homogeneity in prices and in income; in the Hicksian system, we had homogeneity just in prices. "Adding up" for the Marshallian system comes right off the budget constraint; that is, because $\sum_{i=1}^{N} X_i P_i = M$, this must hold at the optimum, so that $\sum_{i=1}^{N} X_i^M(P_1, \ldots, P_N, M)P_i = M$. This doesn't have a whole lot to do with rationality; we've just said that people spend all their income. This holds for all prices, so we can differentiate with respect to the price of good j to get (see chapter 1 for the full derivation):

$$\sum_{i=1}^{N} s_i \epsilon_{ij}^{M} + s_j = 0.$$

Let's suppose $s_j = 0.1$. Then a 10% increase in the price of good j will be a 1% reduction in my real income. Since I have a 1% reduction in my real income, the formula says that, on average, I have to reduce my consumption of other goods by 1%. Gary Becker (1962) used to love this point: a lot of demand comes right off the budget constraint. Even the law of demand is not that far from the budget constraint taken from this perspective. If a good becomes more expensive, you're poorer; you have to consume less, on average. But where will you consume less? It makes sense to think you will consume less of the good that has become more expensive, which is the law of demand.

Chapter 1 does not look at symmetry for Marshallian demand, but we can get it from Hicksian symmetry (chapter 2) and the Slutsky equation. Symmetry for Hicksian demand is $\dfrac{\partial X_i^{H}}{\partial P_j} = \dfrac{\partial X_j^{H}}{\partial P_i}$. We can use the Slutsky equation to write:

$$
\begin{aligned}
\frac{\partial X_i^{M}}{\partial P_j} &= \frac{\partial X_i^{H}}{\partial P_j} - \frac{\partial X_i^{M}}{\partial M} X_j = \frac{\partial X_j^{H}}{\partial P_i} - \frac{\partial X_i^{M}}{\partial M} X_j \\
&= \frac{\partial X_j^{M}}{\partial P_i} + \frac{\partial X_j^{M}}{\partial M} X_i - \frac{\partial X_i^{M}}{\partial M} X_j \\
&= \frac{\partial X_j^{M}}{\partial P_i} + \frac{\partial X_j^{M}}{\partial M} \frac{M}{X_j} \frac{X_i X_j}{M} - \frac{\partial X_i^{M}}{\partial M} \frac{M}{X_i} \frac{X_j X_i}{M} \\
&= \frac{\partial X_j^{M}}{\partial P_i} + \frac{X_i X_j}{M} (\eta_j - \eta_i).
\end{aligned}
$$

In elasticity format, this Marshallian symmetry is:

$$s_i \epsilon_{ij}^{M} = s_j \epsilon_{ji}^{M} + s_i s_j (\eta_j - \eta_i).$$

Thus, symmetry holds for the Marshallian case when the two goods have equal income elasticities (i.e., $\eta_i = \eta_j$). Recall from chapter 1 that the shares and elasticities are evaluated at a particular set of prices and a particular income. So the adding up, symmetry, homogeneity, and the Slutsky equation are describing the demand system at a particular point.

DEMAND SYSTEM DEGREES OF FREEDOM

To make progress on a lot of problems, it's very important to keep your model simple. Limiting the number of goods helps a lot because, at a particular income and set of prices, an N-good model has N expenditure shares, N income elasticities, and N^2 Marshallian price elasticities. These $N(N+2)$ parameters are interrelated because of symmetry, the budget constraint on shares, the budget constraint on income elasticities (i.e., Engel aggregation), and adding up; but those are only $(N-1)N/2+N+2$ restrictions, so the N-good model still has $(N+4)(N-1)/2$ free parameters.[1] For $N=2, 3, 4, 5$, that means 3, 7, 12, and 18 free parameters, respectively.

Take the two-good case. If we are given values for, say, s_1, η_1, and ε_{11}^M, then we can use the budget constraint, adding up, and symmetry to obtain the other five parameters $(\varepsilon_{12}^M, \varepsilon_{21}^M, \varepsilon_{22}^M, s_2,$ and $\eta_2)$. Specifically, the budget constraint gives us the second share from the first: $s_2 = 1 - s_1$. With these shares, the budget constraint also gives us the second income elasticity from the first:

$$\eta_2 = \frac{1 - s_1 \eta_1}{s_2}.$$

Homogeneity gives us ε_{12}^M from ε_{11}^M and the income elasticity:

$$\varepsilon_{12}^M = -\varepsilon_{11}^M - \eta_1.$$

Symmetry gives us ε_{21}^M from ε_{12}^M, the income elasticities, and the shares:

$$\epsilon_{21}^M = \frac{s_1 \epsilon_{12}^M + s_2 s_1 (\eta_1 - \eta_2)}{s_2}.$$

Homogeneity gives us ε_{22}^M from ε_{21}^M and the income elasticity:

$$\varepsilon_{22}^M = -\varepsilon_{21}^M - \eta_2.$$

In the three-good case, we need two shares and two income elasticities and from these and the budget constraint can determine the remaining share and the remaining income elasticity. We also need three cross-price elasticities, such as $\varepsilon_{21}^M, \varepsilon_{31}^M, \varepsilon_{32}^M$, from which symmetry and homogeneity determine the remaining six price elasticities.

The two-good model is one way to keep the model simple. Other times we use the three-good model with additional restrictions. Chapter 8's

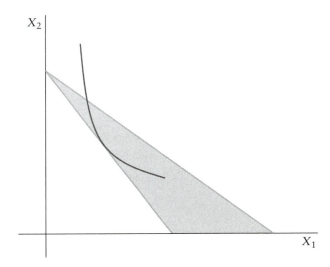

Figure 3-3: A decrease in the price of good 1 creates more opportunities for consumption of good 1 than consumption of good 2.

rent gradient model is an example where two goods—work time and commuting time—are assumed to be perfect substitutes. Chapter 10 introduces production and thereafter the constant returns restriction is frequently referenced. A number of policy questions inherently refer to more than two goods, such as the effects of the corporate-income tax on noncorporate business activity (chapter 18). Here, and for the analysis of optimization over time (chapter 17), the idea of a composite commodity is helpful. All of these special cases of the three-good model are relatively easy to work with because there are fewer than seven free parameters.

THE INCOME EFFECT OF A PRICE CHANGE

Another point Becker (1962) made is about how purchasing opportunities shift when prices change. Suppose the price of good 1 goes down, as in Figure 3-3. The new opportunities will mostly arise for the bundles where a lot of the cheaper good is consumed. After the price is reduced, the "center of gravity," in some sense, moves toward good 1; that is, if I pick a bundle within the budget set at random, it will contain more of good 1 on average.

The income effects discussed so far are at the individual level, but they are informative about aggregate cases too. See Figure 3-4. Different

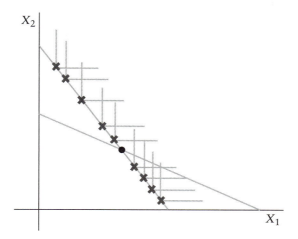

Figure 3-4: An aggregate compensated change (real incomes stay constant) redistributes income from the people who consume lots of X_2 to the people who consume lots of X_1.

people make different choices along a given budget constraint. Consider the average choice. Now suppose prices change in such a way that we pivot around the average point; one price goes up, and one goes down. Those who were consuming more of the good that is now more expensive may not like the new prices. More generally, even though incomes are constant on average, they're being redistributed from the group consuming more of the good whose price increased—in the figure, the redistribution transfers income from the heavy X_2 consumers to the heavy X_1 consumers. Now suppose further that indifference curves for individuals are perfect complements, so individuals do not substitute. On aggregate, we still get substitution because effectively we make the people who like X_1 richer and the people who like X_2 poorer.

Conceivably, people could have enough X_1 so that when it becomes cheaper, they spend all their additional money on X_2. There's a famous theorem by Hugo Sonnenschein that says that because of these types of income effects, you can get just about anything in the aggregate; that is, "anything goes." Draw a crazy curve—then this curve could be a demand curve, in the aggregate. While that can happen, the most common way that income effects work is just to reinforce the story; that is, most frequently they work in the same way as substitution to reinforce the law of demand. See Figure 3-5 for an illustration of how one of these more nonstandard demand curves might result.

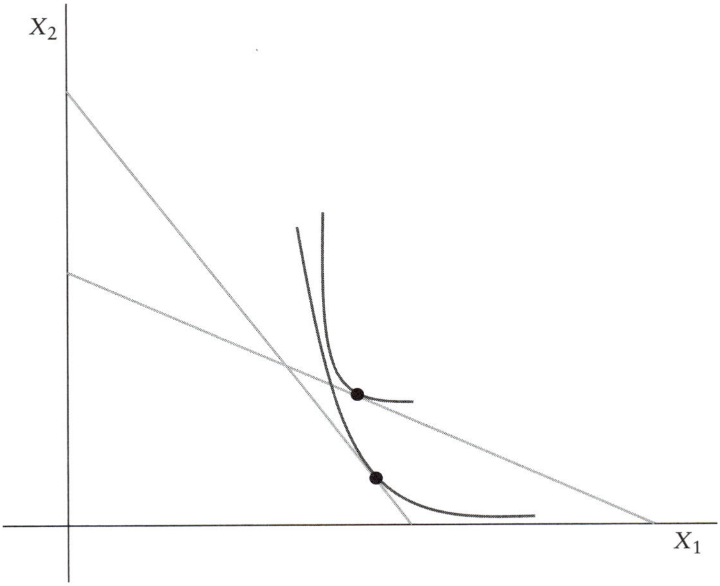

Figure 3-5: After the shift in relative prices, the heavy X_1 consumers buy less of X_1.

To think about these nonstandard cases, consider aggregating the Slutsky equation and then multiplying the income effects term by $\dfrac{\Sigma_{people}\, X_j}{\Sigma_{people}\, X_j}$:

$$\Sigma_{people}\, \frac{\partial X_i^M}{\partial P_j} = \Sigma_{people}\, \frac{\partial X_i^H}{\partial P_j} - \left(\Sigma_{people}\, X_j\, \frac{\partial X_i}{\partial M}\right)\left(\frac{\Sigma_{people}\, X_j}{\Sigma_{people}\, X_j}\right)$$

$$= \Sigma_{people}\, \frac{\partial X_i^H}{\partial P_j} - \left(\Sigma_{people}\, X_j\right)\left(\Sigma_{people}\, \frac{X_j}{\Sigma_{people}\, X_j}\, \frac{\partial X_i}{\partial M}\right).$$

The Marshallian and Hicksian price terms have a natural aggregate interpretation: what happens to aggregate purchases of good i when every person faces the same increase in the price of good j. The aggregate Slutsky equation's final term has aggregate good j consumption ($\Sigma_{people}\, X_j$) exactly where the individual equation had individual consumption. However, consumption is multiplied by an effect that depends on which good's price changes, whereas the individual Slutsky equation would have $\partial X_i/\partial M$, which is properly called an income effect because it is independent of the source of the income (i.e., good j does not appear). Unless everyone has the same income effect, in which case $\partial X_i/\partial M$ can be factored out of the sum across people, it makes little sense to refer to an aggregate income effect because it depends on who gets the income.

From now on, we draw Marshallian demand curves as downward sloping. That is, we rule out the Giffen good case, which means either that the income effect is in the same direction as the substitution effect or that the income effect is sufficiently small.

Chapter 4

Price Indices

Consumer Theory Guides Measurement

Consumer theory gives us a lot of guidance about how to measure things like real income and GDP; it suggests weighting changes in quantities by prices (more expensive goods are more valuable, so changes in their quantities matter more) and weighting changes in prices by quantities (consumers are more affected by changes in the prices of goods that they buy more). We will look now at several of these weighting schemes, motivated by the cost function we studied in chapter 2.

LASPEYRES AND PAASCHE DECOMPOSITIONS OF EXPENDITURE GROWTH

Expenditure is the cost of the chosen bundle of goods—the sum of prices times quantities. We can decompose expenditure growth E_{t+1}/E_t into a price index P_{t+1}/P_t and a quantity index Q_{t+1}/Q_t:

$$\frac{E_{t+1}}{E_t} = \frac{\sum_i X_{i,t+1} P_{i,t+1}}{\sum_i X_{i,t} P_{i,t}} = \frac{\sum_i X_{i,t} P_{i,t+1}}{\sum_i X_{i,t} P_{i,t}} \times \frac{\sum_i X_{i,t+1} P_{i,t+1}}{\sum_i X_{i,t} P_{i,t+1}} = \frac{P_{t+1}}{P_t} \times \frac{Q_{t+1}}{Q_t}.$$

It's especially useful because often we only measure two of the three. Price is sometimes easier to measure than quantity because price can be seen from a sample; just go to one of the sellers in the market—say, a grocery store and look at its price for eggs. Quantity measurement can be more difficult—you have to ask every seller what they sold; you'd need some kind of census of sellers. In these situations we often back out

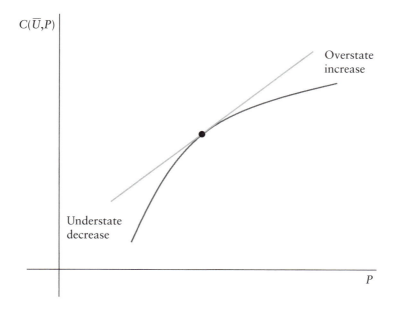

Figure 4-1: Because of the concavity of the cost function, an increase in the cost of living is overstated, and a decrease in the cost of living is understated.

quantities by measuring expenditure and price and using the decomposition above.

To make the decomposition, we simply multiplied $\dfrac{\sum_i X_{i,t+1} P_{i,t+1}}{\sum_i X_{i,t} P_{i,t}}$ by $\dfrac{\sum_i X_{i,t} P_{i,t+1}}{\sum_i X_{i,t} P_{i,t+1}} = 1.$. But why does this make sense? So long as the numerator and denominator are the same, we could multiply by any number of things to decompose $\dfrac{\sum_i X_{i,t+1} P_{i,t+1}}{\sum_i X_{i,t} P_{i,t}}$. The first part of the answer is that $\dfrac{\sum_i X_{i,t} P_{i,t+1}}{\sum_i X_{i,t} P_{i,t}}$ is a first-order approximation to the cost function. That is, this is saying how much the cost increased along the tangent line; it tells us how much it would cost us to buy today's bundle tomorrow relative to today. Consider Figure 4-1. Because the line is a first-order approximation, we overstate an increase in the cost of living and understate a decrease in the cost of living due to the concavity of the cost function.

The second part of the answer is that $\dfrac{\sum_i X_{i,t+1} P_{i,t+1}}{\sum_i X_{i,t} P_{i,t+1}}$ is a quantity index that approximates the indifference curve using the budget line. This makes sense because the consumer's indifference curve is tangent

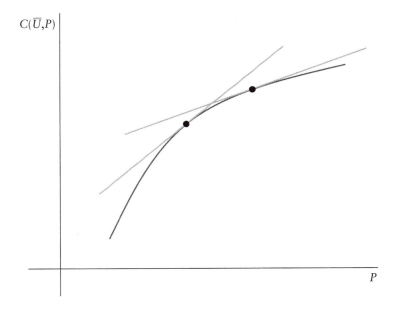

Figure 4-2: A cost function approximated at two different points. Because the cost function is concave, depending on where we move from t to $t+1$, one estimation is too large and one is too small. These correspond to the Paasche and Laspeyres price indices.

to the budget line. Thus, changes in utility can be approximated by movements along the budget line. This is again, of course, a first-order approximation, and it is measured in dollars; more specifically, it tells us the income equivalent required to give you the change in utility.

This case was for a price index based in period t and a quantity index based in $t+1$, since prices were weighted by period t quantities and quantities were weighted by period $t+1$ prices. Again, because our choices of numerator and denominator were arbitrary, we could do this the opposite way. Consider decomposing expenditure growth into a price index based in $t+1$ and a quantity index based in t:

$$\frac{E_{t+1}}{E_t} = \frac{\sum_i X_{i,t+1} P_{i,t+1}}{\sum_i X_{i,t} P_{i,t}} = \frac{\sum_i X_{i,t+1} P_{i,t+1}}{\sum_i X_{i,t+1} P_{i,t}} \times \frac{\sum_i X_{i,t+1} P_{i,t}}{\sum_i X_{i,t} P_{i,t}} = \frac{P_{t+1}}{P_t} \times \frac{Q_{t+1}}{Q_t}.$$

In general, one decomposition is not better than another. See Figure 4-2. The different methods simply change where the first-order approximations occur—either in period t or period $t+1$.

The first price index we considered, based in period t (i.e., prices are weighted by period t quantities), is called the Laspeyres price index. The second case we considered, based in period $t+1$, is called the Paasche price index.

These measures are sometimes combined as a geometric average in an index called the Fisher ideal index. For prices, the Fisher ideal index is:

$$\left(\frac{\sum_i X_{i,t} P_{i,t+1}}{\sum_i X_{i,t} P_{i,t}} \right)^{\frac{1}{2}} \left(\frac{\sum_i X_{i,t+1} P_{i,t+1}}{\sum_i X_{i,t+1} P_{i,t}} \right)^{\frac{1}{2}}.$$

Unlike the Laspeyres and Paasche indices, we now no longer know the direction of the bias (as a measure of the movement along the cost function from t prices to $t+1$ prices). On the other hand, we know that this measure will be better than at least one of the Laspeyres and Paasche indices (i.e., less biased).

CHAINED PRICE INDICES

Suppose we want to perform this calculation over a long period of time. We can consider the two Laspeyres and Paasche ratios for looking at this:

$$\frac{\sum_i X_{i,1950} P_{i,2015}}{\sum_i X_{i,1950} P_{i,1950}} \quad \text{versus} \quad \frac{\sum_i X_{i,2015} P_{i,2015}}{\sum_i X_{i,2015} P_{i,1950}}.$$

These will give such radically different answers that they are essentially useless. Consider the denominator on the right term, and take i to be cell phones. What is the price of a cell phone in 1950? It would be absurdly high. A single modern phone probably does more computing than the aggregation of all computing prior to 1950. You'd never consider buying your 2015 bundle at 1950 prices. Similarly, consider the numerator on the left side. We'd leave out all the new goods that weren't around in 1950. Cell phones, for instance, would get far too little weight. In short, ignoring the substitution effect introduces large errors over long time horizons. The way we address this is with a chained price index. Consider the following:

$$\frac{P_{2015}}{P_{1950}} = \left(\frac{P_{1951}}{P_{1950}} \right) \left(\frac{P_{1952}}{P_{1951}} \right) \cdots \left(\frac{P_{2015}}{P_{2014}} \right).$$

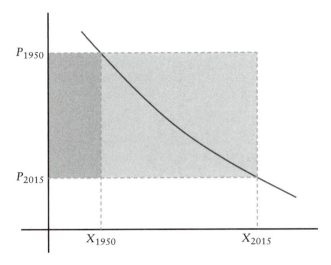

Figure 4-3: The change in the cost of living is vastly underapproximated or overapproximated by the two naïve attempts to measure it.

That is, the change in the cost of living between 1950 and 2015 was really a series of changes going from one year to the next. Then we can write all of these in terms of price indices:

$$\left(\frac{\sum_i X_{i,1950}\, P_{i,1951}}{\sum_i X_{i,1950}\, P_{i,1950}}\right)\left(\frac{\sum_i X_{i,1951}P_{i,1952}}{\sum_i X_{i,1951}\, P_{i,1951}}\right)\cdots\left(\frac{\sum_i X_{i,2014}\, P_{i,2015}}{\sum_i X_{i,2014}\, P_{i,2014}}\right).$$

In this formulation, new goods will be added into the price indices as they arrive. Cell phones will be added in when they become cheap enough that people will actually buy them.

Now let's visualize our naïve approaches to measuring the change in the cost of living between 1950 and 2015. Expressed in dollars rather than ratios, the Laspeyres and Paasche approaches are $\sum_i X_{i,1950}\,(P_{i,2015}-P_{i,1950})$ and $\sum_i X_{i,2015}\,(P_{i,2015}-P_{i,1950})$. Let's assume that only one price changed, and that income is the same in the two years, so that we can draw the changes in the usual Marshallian demand picture, as in Figure 4-3.

The smaller rectangle estimates the change in the cost of living the same way that the Laspeyres index does: using the initial consumption bundle to weight the price changes. The larger rectangle uses the final consumption bundle, as the Paasche index does.

The true change in the cost of living that the two indices are designed to approximate is

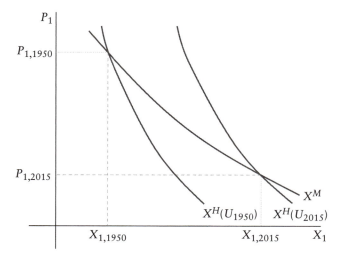

Figure 4-4: Hicksian demand curves tell the answers to two different questions: how much cheaper is it to get the 1950 level of utility, or how much more expensive is it to get the 2015 level of utility?

$$C(P_{1,t+1}, \ldots, P_{N,t+1}, \overline{U}) - C(P_{1,t}, \ldots, P_{N,t}, \overline{U}).$$

We still have the same issue of a reference point. The ambiguity this time is about which $\overline{U}$ to use (period t utility or period $t+1$ utility?). Now, however, the issue is not about what approximation to use; changing which $\overline{U}$ we use corresponds to different questions. We may want to know how much it costs to get 2015 utility in 1950, or we may want to know how much it costs to get 1950 utility in 2015.

Now, for simplicity, as in Figure 4-3, assume only the price of good 1 is changing. Then we can write

$$C(P_{1,t+1}, \overline{P}_2, \ldots, \overline{P}_N, \overline{U}) - C(P_{1,t}, \overline{P}_2, \ldots, \overline{P}_N, \overline{U})$$
$$= \int_{P_{1,t}}^{P_{1,t+1}} \frac{\partial C}{\partial P_1} \, dP_1 = \int_{P_{1t}}^{P_{1,t+1}} x^H(P_1, \overline{P}_2, \ldots, \overline{P}_N, \overline{U}) \, dP_1.$$

Because the change in cost is an area to the left of a Hicksian demand curve, Figure 4-4 adds some Hicksian demand curves to Figure 4-3, assuming that good 1 is a normal good, which makes the Hicksian curves less price sensitive than the Marshallian curve.

In Figure 4-4, we drew the Hicksian demand curves much steeper than the Marshallian demand curves. This is useful for visualization, but a

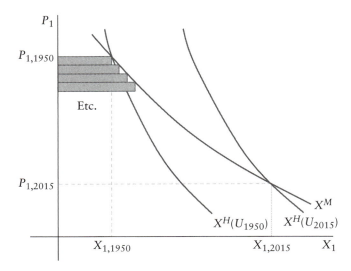

Figure 4-5: The chained price index roughly gives us the area under the Marshallian demand curve.

more realistic picture would show less contrast between the slopes. Recall the Slutsky equation. If the income elasticity is 1, and the share of the good is 1%, then Marshallian and Hicksian elasticities will differ only by .01. Keep this in mind when thinking about how the Marshallian and Hicksian demand curves are related.

But what about using the Marshallian demand curve to measure the cost of living? Looking at Figure 4-4, it doesn't seem so crazy, because the area under the Marshallian curve gives us something in between the areas under the two Hicksian curves.

It should be noted, however, that the two Hicksian demand curves in Figure 4-4 are answering two slightly different questions. The Hicksian demand curve on the left answers how much cheaper it is to attain the 1950 level of utility, while the Hicksian demand curve on the right indicates how much more expensive it is to attain the 2015 level of utility.

A chained price index estimates a cost change from 1950 to 1951, adds a cost change from 1951 to 1952, and so forth. Each link in the chain is weighting price changes by quantities that people were buying at that time: namely, the quantity on the Marshallian demand curve. The chained price index can therefore be visualized in the same picture: see Figure 4-5.

The Fisher index, on the other hand, will simply average the Laspeyres and Paasche indices that we've already depicted in Figure 4-3. Consider Figure 4-6.

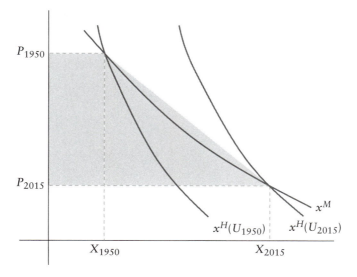

Figure 4-6: The Fisher index averages the Laspeyres and Paasche indices.

Remember the problem motivating our use of price indices—we want to figure out how important price changes are over time. The demand curve gives us a lot of information about this. Think again about the one-good case and consider Figure 4-7. The marginal value of an additional unit has to be equal to the price I'm paying. So we don't ask people how important a good is—people reveal how important it is by how much they're willing to consume when the price is higher than it is today.

Even with the chained index, we only account for new goods with some lag. That is, there is a period in which people are buying them but we haven't incorporated them into our index yet. Then we miss the initial benefit when the price was high. See Figure 4-8.

Ideally, the good's price would be recorded as the choke price during periods when the good is unavailable, so that the chained index would capture Figure 4-8's shaded area. But in typical practice, we measure prices only when a good is traded.

USING THE COST FUNCTION TO VALUE QUALITY CHANGE

The other issue with these price indices is that goods are not the same over time; quality changes. One method to account for quality is by using

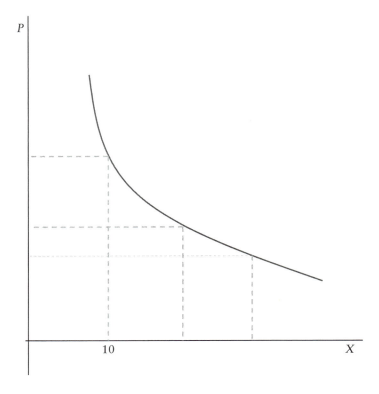

Figure 4-7: The marginal value of the 10th unit must be equal to the price. So people reveal how much the 10th unit is worth to them.

components. With a car, for instance, we might keep track of miles per gallon over time as one measure of the car's quality. However, a simple model that many economists use is just that higher quality means a lower effective price. Then one unit of today's good equals K units of the old good. Suppose X_i is a good that got better, so that there is an old X_i and a new X_i. Then

$$X_i^{new}(P_1, \ldots, P_{i-1}, P_i, P_{i+1}, \ldots, P_N, M)$$
$$= \frac{1}{K} X_i^{old}(P_1, \ldots, P_{i-1}, P_i/K, P_{i+1}, \ldots, P_N, M)$$

because the per-unit price of good i is scaled by $1/K$, and the old units only require $1/K$ new units. Think about it this way. Candy bars used to cost \$0.50 for 1 oz. Now, they're \$0.50 for 2 oz. The price per ounce has moved from \$0.50 to \$0.25. It's obvious that we want more ounces after

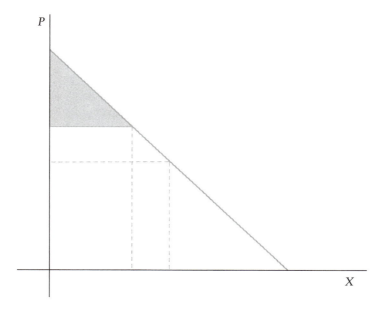

Figure 4-8: When we incorporate new goods with some lag, we miss the shaded region in the graph.

the bars got bigger, but do we want more bars? Suppose the elasticity of demand is −1. Then we demand twice as many ounces after the price is cut in half, and this means the same number of bars. Thus, if demand is elastic or inelastic, we demand more or fewer bars, respectively. Therefore, an increase in quality might reduce demand (for total quantity of units). This is realistic. If tire quality increases, for instance, we replace our tires less frequently.

We can also use this example about tires to think again about the demand elasticity. Why do we think that the demand for tires is inelastic, for instance? Suppose the price of tire-miles falls so that it is half of what it used to be. How much has the price of driving fallen? It's clear that the amount is much less. So even if demand for driving is elastic, the fall in the price of tires does not reduce the total price of driving by much, since tires are a small share of driving costs. And thus the increase in demand for driving will not increase the demand for tires by very much. Now, what about a single manufacturer of tires? Suppose a single manufacturer reduces the price of its tires. We know the demand will be elastic here. Why? If demand were inelastic, the manufacturer would *raise* prices to make more money.

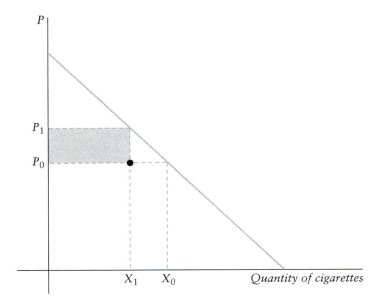

Figure 4-9: We can value people's change in views about the health effects of cigarettes in response to new information by the shaded region above.

Suppose our econometrics skills are bad, so that we can't estimate the elasticity of demand for X_i as in the equation above. An increase in quality still has a very predictable effect on complements and substitutes:

$$X_j = X_j(P_1, \ldots, P_{i-1}, P_i/K, P_{i+1}, \ldots, P_N, M).$$

That is, if good i gets better, people will buy fewer of the substitutes for i and more of the complements to i.

Now suppose we know the demand function for a good. Take cigarettes, for instance. Assume people learn that cigarettes are harmful and, in response, decrease consumption from X_0 to X_1 while the monetary price of cigarettes remains P_0. How much did people update their assessment of the health cost of smoking? Well, how much would we have had to raise the price of cigarettes to get people to reduce their consumption to X_1? The shaded region in Figure 4-9 tells us the answer to this: on a per-unit-quantity basis, people increased their assessment of the health costs of smoking by $P_1 - P_0$.

They are acting like the price of cigarettes is P_1, which is the sum of the monetary cost P_0 and a perceived health cost $P_1 - P_0$.

Chapter 5

Nudges in Consumer Theory

INDIFFERENCE CURVES FOR BUYERS

Recall that an individual's demand curve tells us, for a given price, what quantity he or she chooses to buy. Then, for a given price, any point to the right or left of this point leaves the consumer worse off. Further, for a given quantity, lower prices are always preferred. Using this intuition, we can draw an indifference curve together with a (Marshallian) demand curve in demand space. See Figure 5-1.

By including the indifference curve, which is tangent to a horizontal price line where it crosses the demand curve, Figure 5-1 makes it more obvious that any point on a demand curve represents the consumer's optimal quantity to purchase given the price being charged.

In general, what do we think about someone making a purchase to the right of their demand curve? Is this a puzzle for consumer theory? No, because on the margin it is not often worth the effort to get it exactly right. Notice, in Figure 5-1, that the slope of the indifference curve is zero at the demand curve, which means that a consumer is essentially indifferent between the point on his or her demand curve and points slightly to the right of that demand curve. There is only the second-order effect coming from the curvature of the indifference curve that makes the consumer worse off. So, we're not normally worried if people purchase a few more units at a given price than is optimal for them.

What if, however, it's a matter of someone being *above* their demand curve, in the sense of Figure 5-2 below? This is a consumer paying more for a given bundle than that bundle would warrant on the demand curve.

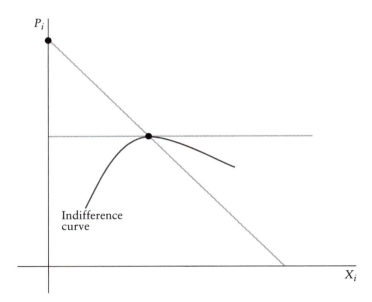

Figure 5-1: In (X_i, P_i) space, we can draw indifference curves. Each indifference curve is tangent to a price line where it crosses the demand curve, which indicates the optimal quantity to purchase at a given price.

Here, consumer theory is contradicted. The cost of paying a higher price for a given bundle on the demand curve is much higher than the cost of buying more units than demand would suggest at a given price. What the consumer loses from paying a higher price is depicted by the shaded region in Figure 5-2.

People shop this way at grocery stores. They shop around intensively on price, but once they're in the store, they frequently buy more than their demand curve would suggest at given prices.

CONSUMER MISINFORMATION AND "NUDGEABILITY" IS A PREDICTION OF CONSUMER THEORY

Consumers have a strong incentive to pay less for a given quantity, but little incentive to purchase exactly the right quantity at a given price. Consumer theory is telling us that people should be pretty open to suggestions as to how much to buy, as long as it is not too far to the right or left of their demand curve. It is also telling us that, to the extent that information is costly, people will be somewhat misinformed about the quantity they should be purchasing. Such misinformation is causing them

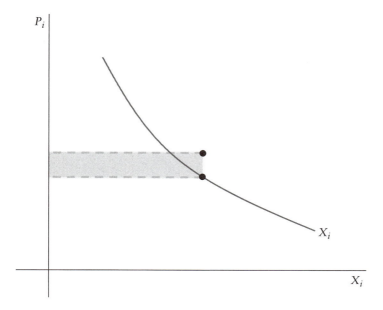

Figure 5-2: We care about going to a supermarket that charges more than another, but we're not too concerned about Coke versus Pepsi.

to buy too much or too little, but those purchase mistakes are hardly affecting consumer welfare as long as they are not too large. Why pay for information that has little value?

A growing literature in economics asserts that consumers are sometimes misinformed about what they buy and will make different choices if nudged to do so (Thaler and Sunstein 2008). The observation is not surprising, but we are surprised when the authors of such studies assert that consumer theory has been refuted. Consumer misinformation and responsiveness to suggestion are predictions of consumer theory, as long as the acquisition of information is not free.

Chapter 13 revisits this in the market context, showing that the "nudge-ability" of consumers has many times been a force promoting competition and thereby enhancing efficiency and consumer welfare.

Chapter 6

Short- and Long-Run Demand, with an Application to Addiction

AN EXAMPLE: THE DEMAND FOR CARS AND GASOLINE

To illustrate the distinction between long-run and short-run demand, take the demand for gasoline. Let's begin with Figure 6-1.

The market starts with price and quantity P_0, Q_0. What would happen if the price of gasoline instead moved up to P_1 (a dramatic increase, as we have seen numerous times in history)? The law of demand tells us that people buy less gas. But we think that the demand for gasoline, at least in the short run, is going to be relatively inelastic. If I double the price of gasoline, I might see the quantity of gas go down maybe 10% or so, such as the quantity Q_1 in the figure below.

Many people say that the demand for gasoline is completely inelastic, which is incorrect; people do save gas when the price increases. What are the major ways in which people can conserve on gas? People take fewer or shorter road trips; they go on a shorter vacation or they don't drive as far on a vacation as they otherwise would have done. People might also take the bus or the subway, walk, or do something else other than drive. All of these are ways of "driving less." As when we talked about cars and tires, the activity that is being adjusted is the amount of driving. If we were going to achieve a 10% decrease in gas consumption, do we have to drive 10% fewer miles? Is driving fewer miles the only way we could consume less gas? Many households have more than one car; that's one thing that's changed over time. People can drive a smaller car and leave the minivan at home. They can still take the same trip, but use the smaller, more fuel-efficient car. People won't change the composition of cars they

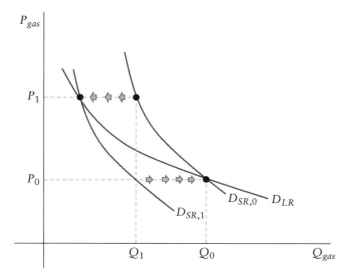

Figure 6-1: The long- and short-run demand for gasoline. Along long-run demand, all other prices are held constant, and all other quantities adjust, especially the quantity of cars. Each short-run demand curve corresponds to a specific quantity of cars.

own overnight, but they can still change the composition of cars on the road. It may take time to adjust the car fleet that is in the garage but the car fleet on the road can change instantly. This short-run effect of changing which cars are driven has become much more important as we have more and more multi-car households. Nonetheless, in the short run there's a limited number of possible adjustments. That's why the short-run demand curve $D_{SR,0}$ is relatively steep.

Now what would happen if the price of gas stayed up at P_1? People would buy smaller or more fuel-efficient cars. When the price of gas goes up, sales of pickup trucks go down and sales of more fuel-efficient cars go up. There are other margins for adjustment. Cities could build public transportation or people could move closer to work—there are many more possible responses in the long run than in the short run. So, over time the consumption point moves to the left of Q_1. The response to the price increase is more elastic in the long run because we have more substitution. In Figure 6-1, the long-run demand curve D_{LR} is flatter than the short-run demand curve $D_{SR,0}$.

Even though we are discussing consumer demand, which is often derived from utility, we see that consumer demand is sometimes more easily understood in terms of production. One of the major shifts making long-run demand for gas different from short-run demand is that the car fleet is changing—that's the production function approach. Consider the two points in Figure 6-1 at price level P_1. The car fleet looks different at the right point than at the left point. Think about the 1970s in the United States, for instance, when the price of gas rose a lot; if you look at the fleet of cars, it changed dramatically. The 8-cylinder car went out of production by the 1980s.

What would happen if, when we arrived at the long-run demand curve, the left point at price level P_1, the price fell back to where it used to be, P_0? We would move along another short-run demand curve, $D_{SR,1}$. This curve would be again less elastic because the fleet of cars is being held constant along that curve. Then what would happen? It was amazing how far back we actually came. When gas prices went up, cars became much more fuel efficient; then gas prices went down and what happened? The 8-cylinder car came back; even 10-cylinder cars came back. People have a strong taste for these cars. Technology did change; engines are much more efficient than they used to be, but what do we use that efficiency for? More power and more performance. Performance of cars has gone up enormously: a Honda Civic is a high-performing car by prior standards. We have an elastic demand for performance, and as the price of gas decreased we used a lot of the more fuel-efficient technology for better performance rather than for saving gas.

The key notion is that there's a lot more adjustment that happens in the long run that cannot happen in the short run. This is important for many goods, because people can respond more as they have more and more things they can change, and cars and gas are good examples of that. We look at another example later in this chapter.

RELATING THE SHORT-RUN DEMAND CURVE TO THE OVERALL DEMAND SYSTEM

In the short run, the quantity of cars is held constant, whereas a demand system reflects the assumption that all quantities adjust:

$$X_G = X_{Gas}\,(P_{Gas}, P_{Cars}, \ldots)$$
$$X_C = X_{Cars}\,(P_{Gas}, P_{Cars}, \ldots)$$

where we are denoting quantities with X, prices with P, and specific goods with subscripts. Each quantity depends on all of the prices.

The demand system says that the demand for gas depends on, among other things, the price of gas and the price of cars. Obviously, we expect the first effect to be negative, but the second effect is also negative because cars and gas are complements. When cars are cheaper, we will want more cars and more gas.

To get gasoline's long-run demand curve, we take the demand system and vary the price of gas, holding all other prices constant:

$$\frac{dX_G^{LR}}{dP_G} = \frac{\partial X_{Gas}}{\partial P_G}.$$

The long-run response looks at how the quantity of gas responds to its price when people can freely adjust their other goods and, importantly, freely adjust what car they own (or whether they own a car at all).

Now, let's use the same demand system to get the short-run demand curve. What is different about the short run is that people cannot change the quantity of cars. This is not a law; but the supply of cars is fixed in the short run, so prices—especially the price of cars—must adjust in the short run so that people are willing to buy those cars:

$$dX_G^{SR} = \frac{\partial X_{Gas}}{\partial P_G} dP_G + \frac{\partial X_{Gas}}{\partial P_C} dP_C.$$

The first term is the long-run effect and the second term reflects the short-run change in the price of cars. Moreover, the demand system tells us the amount dP_C by which car prices have to change so that when gas becomes expensive, people are still willing to have their cars:

$$0 = dX_C^{SR} = \frac{\partial X_{Cars}}{\partial P_G} dP_G + \frac{\partial X_{Cars}}{\partial P_C} dP_C.$$

The zero is the change in supply, which has to equal the change in demand. We solve this for the car price change that has to occur with each unit of gas price change in order that the quantity of cars remains unchanged:

$$\frac{dP_C}{dP_G} = -\frac{\dfrac{\partial X_{Cars}}{\partial P_G}}{\dfrac{\partial X_{Cars}}{\partial P_C}}.$$

If cars and gas are complements, the cross-price term (the numerator) is negative; as the price of gas goes up, I want fewer cars. The entire term is therefore negative, which implies that the price of cars is going to have to go down to hold the stock of cars that people have constant. We can substitute this equation into the gas-demand equation and get

$$\frac{dX_G^{SR}}{dP_G} = \frac{\partial X_{Gas}}{\partial P_G} - \frac{\dfrac{\partial X_{Gas}}{\partial P_C} \dfrac{\partial X_{Cars}}{\partial P_G}}{\dfrac{\partial X_{Cars}}{\partial P_C}}.$$

Remember that the first term is negative because it is the own-price long-run effect. And the sign of the second term, including the minus sign, will be positive. This means that the short-run effect will be smaller than the long-run effect because it will be muted by that second term. As the price of gas goes up, people will want to buy fewer cars, but the price of cars will go down. This will cause people to hold more cars in the short run than they will in the long run, which makes them buy more gas in the short run than they will in the long run. This is the basic mechanism by which the short-run demand will be less elastic than the long-run demand.

In this example, cars and gas are complements, but in other applications we are interested in substitutes. With substitutes, both of the two cross-price terms in the numerator are negative. But their product is still positive, so regardless of whether we have substitutes or complements, the long-run response is greater than the short-run response. The difference is in the mechanism: make gas more expensive and its substitutes become more expensive in the short run, because by definition their quantities are fixed. On the other hand, its complements, such as cars, get cheaper.

USING CONSUMPTION STOCKS TO UNDERSTAND ADDICTION

We have examined the market's inventory of cars as an example where elements of consumption cannot adjust in the short run, thereby creating different long- and short-run demands for complements such as gasoline. Here we show how habits and addictions are also examples.

What is the key aspect of addiction? That is, what is the key feature of demand for addictive goods that differs from what we have discussed in the context of the classical demand model? We start by noting that past consumption will matter for consumption decisions today. This is a particular type of complementarity, namely between past and present. The model in Becker and Murphy's (1988) paper on addiction represents complementarity over time with the standard perpetual-inventory formula used for production stocks, except now it is used for a "consumption stock":

$$S_t = (1 - \delta)S_{t-1} + C_t$$

with the depreciation parameter $\delta \in (0,1]$. The stock-evolution equation can be solved backwards to get the stock as a function of the entire consumption history

$$S_t = \sum_{j=0}^{T}(1-\delta)^j C_{t-j}$$

where date $t - T$ is the first time that the consumer consumed this good. Here we have no algebraic difference from production stocks. The practical differences are that this stock S is not tangible in the way that a stock of houses or vehicles would be, and that additions to the stock are referred to as "consumption" rather than "investment." S represents the consumer's historical experiences. Those are important for habits and addictions because your willingness to pay today depends on how much you have consumed in the past.

Then consumers will solve

$$\max v(y) + \sum_{t=0}^{\infty} \frac{1}{(1+\rho)^t} U(C_t, S_{t-1})$$

where y summarizes the consumption of all other goods that are not habit forming. At this point, we could replace each stock term with its backward-looking consumption expression and note that the objective being maximized here is "just" a function of consumption at each point in time, fitting into the general consumer framework introduced in chapter 1. But, as we always do when applying consumer theory, we take the simplest case that represents the phenomenon at hand, which here is habitual behavior. So we have added only one complexity to our previous setup, which is that utility depends on the consumption-stock variable.

We will assume $U_{CS}>0$, so that I get more utility on the margin from consumption as I increase my stock of consumption of that good. We will assume that U_{CS} is positive enough that consumption at one point in time reinforces consumption in the future (especially the near future).

We know $U_C>0$. But what about U_S? We will denote $U_S<0$ as a harmful addiction and $U_S>0$ as a beneficial addiction. Exercise might be considered a beneficial addiction, for example. But the distinction between harmful and beneficial is not as important as our assumption that current consumption reinforces future consumption. Current and future consumption are complements. The complementarity is a common element between exercising, listening to classical music, or doing price theory (!), on the one hand, and smoking cigarettes or taking cocaine on the other.

For the same reason, the fact of complementarity (requiring that $U_{CS}>0$) does not tell us whether the complementary activity is a good ($U_S>0$) or bad ($U_S<0$). This is sometimes a source of confusion in social interactions research. When my neighbor buys a faster car, I respond by buying a faster car. That tells us that his consumption of fast cars is complementary to my consumption of fast cars. That does *not* tell us that his purchase harmed me, even though the result of his purchase is that I spend more money on cars. A better way to assess whether one person's fast car harms or helps his friends is to look at how people choose their friends. Do they choose friends with slow cars so that they can feel better?

Now think about the marginal utility received from an additional unit of consumption at time t:

$$\frac{1}{(1+\rho)^t}U_C(C_t,S_{t-1}) + \frac{1}{(1+\rho)^{t+1}}U_S(C_{t+1},S_t) + \frac{1-\delta}{(1+\rho)^{t+2}}U_S(C_{t+2},S_{t+1})+\cdots$$

$$= \frac{1}{(1+\rho)^t}\left[U_C(C_t,S_{t-1}) + \frac{1}{1+\rho}\sum_{j=1}^{\infty}\left(\frac{1-\delta}{1+\rho}\right)^{j-1}U_S(C_{t+j},S_{t+j-1})\right].$$

The first term is the usual term—more consumption today is valuable from today's perspective. The rest of the terms reflect that period t's consumption adds to the habit stock, especially in the near future. A rational consumer takes the future terms into account in making his decision about period t's consumption. For a harmful addiction, all of the future terms are negative.

We say that consumption is complementary over time when the marginal utility expression above is increasing in the stock S_{t-1}. This is stronger than $U_{CS} > 0$ because S_{t-1} increases the future stocks, which reduces the marginal utility (or increases the marginal disutility) of future stocks.

SHORT- AND LONG-RUN PRICE EFFECTS ON ADDICTIVE BEHAVIORS

As will also be discussed in chapters 15 and 16, we consider here a steady state and how the stock S approaches the steady state along an optimal time path. The steady state has $C = \delta S$. We expect the law of demand to hold: steady-state consumption is less when the steady-state price of C is higher.

The history of consumption matters only through S_{t-1}. That is, S is a state variable much like production capital k is in chapters 15 and 16. Let the policy function $C(S)$ denote the consumer's optimal choice for current consumption when his past is summarized by the stock S. The slope of this function tells us whether the good is addictive—whether consumption is complementary over time. It slopes upward in the addictive case.

Figure 6-2 is a diagram that uses these ideas to understand optimal consumption dynamics. A steady state has to be on a ray from the origin. The policy function slopes upward and may, as in the figure, cross the ray from above (ignore the upper policy function for the moment).

The policy function shows us that a consumer beginning with less than steady-state stock consumes above the ray, which is equivalent to saying that he or she consumes more than the stock's depreciation. The consumer's stock is therefore increasing over time and approaches the steady state from the left. A similar argument says that a consumer beginning with more than steady-state stock approaches the steady state over time from the right. In other words, when the policy function crosses the ray from above, the crossing point is a stable steady state.

Now suppose that the consumer is in a steady state and is surprised by a sudden and permanent decrease in the price of the addictive good. The new policy function is above the old one, as shown in Figure 6-3. Consumption increases in the short run—before the stock can change—to the

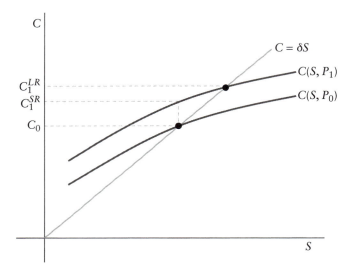

Figure 6-2: When prices are decreased, the $C(S)$ schedule shifts upward and-long run consumption of the addictive good increases. The 45-degree line is the steady state, where consumption simply replaces decreased consumption stock.

point above on the new policy function. But that new consumption level is still below the new steady-state consumption.

In other words, the sudden and permanent price decrease increases consumption more in the long run than in the short run. If you want to know how much price ultimately depresses the consumption of an addictive good, looking at the short-run response is going to provide quite an underestimate.

This is similar to the short-run and long-run effects of a change in the price of gas on consumption of cars and gasoline. A sudden and permanent change in gas prices immediately changes gasoline purchases, without changing the number or types of cars that people own. Over time, cars change too, so that the long-run response of gasoline purchases is greater. Cars are not literally a stock of historical gasoline consumption, but the complementarity between gasoline and cars looks a lot like the complementarity between S and C in our addiction model.

The $C(S)$ schedule might be sufficiently nonlinear, however, that in some range it crosses the ray from below. It may also cross it again from above, as in Figure 6-3. In this case, there is more than one steady state and one of them is unstable. If you start to the right of S^* (the unstable steady-state stock), one will move to the right, and if you start to the left, you move down and eventually quit.

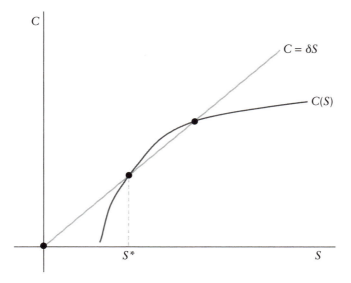

Figure 6-3: Case with an unstable steady state. A perturbation from S^* either leads someone to quit or to the higher level of steady-state consumption.

We can think about rehabilitation in this model. Consider shifting down the $C(S)$ schedule in Figure 6-3. Then someone beginning at S^*, for instance, will begin to move toward quitting along the policy function where it is below the ray. When rehab ends, the policy function returns to the original $C(S)$ curve depicted, but by this time may be at another $C=0=S$ steady state that is stable.

What does this tell us about potential problems caused when addictive substances are declared illegal? When consuming addictive substances is illegal, access to good ways to quit can be reduced. Someone who is addicted to an illegal drug may not be able to take advantage of quitting alternatives without exposure to the risk of prosecution, for example. However, with rational consumers, this is a double-edged sword. The easier it is to quit a harmful addiction, the more people start. Depending on the elasticity, making it easier to quit could actually make *more* people addicted, if more people start than quit.

There are some policies about which we should also be careful. Lying to people, for example, to overinflate how bad a drug is could actually get people to quit. In the long term, however, it detracts from people's trust of those who lied. The liars, in the future, will have less credibility even when they tell the truth. The costs of this are difficult to measure. Furthermore, what about the policy of trying to inform people exactly how bad drugs are? Some studies have shown that people *overestimate* how likely smoking is to cause lung cancer, for instance. We have to take these studies with a grain of salt, however, because we are concerned with choices people make, not choices they say they would make. Fewer studies have been conducted along these lines.

There are also information problems with addictive goods, at least when they are new. It is often difficult to know how addictive a particular good is immediately. There is room for a company putting out a new product to misinform consumers about how addictive it is because their claims are not verifiable for an extended period of time. This is not necessarily unique to addictive goods, however. A dietary supplement could claim to have cancer prevention properties that are realized after twenty years of taking it. This would be very difficult to verify.

Homework Problems for Part I

Prices and Substitution Effects

Chicago price theory has traditionally posed two kinds of homework/ test questions: structured and "TFUs." The structured questions lead the student a bit in setting up a model and introducing notation. The "TFUs" are completely open ended, and students are directed to explain why they think the statement given is "True," "False," or "Uncertain." Both types of questions are about real-world behaviors.

Students are not told which homework goes with which chapters/ lectures. Real-life situations do not come with such labels, so neither do the homework questions. This is a good place for inverting the classroom, so that the instructor can share his or her experience with matching situations to specific price-theory tools.

1) A consumer buys both food, F, and other goods, Y. The consumer's utility depends on the amount of food, F, consumed and on the amount of Y consumed (i.e., $U = U(F, Y)$). Set the price of good Y to 1 (so that all prices are measured relative to the price of Y). Food is a composite good produced using k individual ingredients, $x_1, x_2, \ldots x_k$ according to the production function $F = F(x_1, \ldots x_k)$, where $F(\ )$ has constant returns to scale (i.e., doubling all inputs doubles the amount of food produced).

   a. If you knew the production function F, how would you measure the price of food? How would you measure the quantity of food consumed? If you had to approximate these measures in practice and did not know the function F, what would you do?

b. What will determine the elasticity of demand for a particular ingredient? Will it be larger or smaller than the elasticity of demand for food as a whole?

c. How will a change in the price of one ingredient affect the price of food and the overall consumption of food?

d. For the case of two inputs in the production of food (i.e., $k = 2$), how will the own-price elasticity of demand for the two goods compare? Which good will have a more elastic demand curve? Does it matter if you consider compensated or uncompensated demand elasticities?

2) Now consider the case of "social" goods, like fast cars (in this context we define social goods to be goods where my utility is affected by the consumption choices of others). Utility depends on three things: your consumption of a nonsocial good, Y; your consumption of the social good, X; and your friend's consumption of the social good, X_f, as in $U = U(Y, X, X_f)$. Each person has one friend and the friend has the same preferences with the roles of X and X_f reversed (i.e., the problem is symmetric).

a. Under what conditions on U will an increase in my friend's consumption of fast cars increase my consumption of fast cars?

b. Under what conditions on U will an increase in my friend's consumption of fast cars increase my utility?

c. If, all else being equal, people prefer to have friends with fast cars, what can you infer about preferences?

d. If your friend having a fast car makes you jealous (by which I mean it makes you want to get a fast car too), what does that say about your preferences?

e. For a fixed pairing of individuals (i.e., a fixed set of friends) how would the choice of cars be determined in equilibrium if each acted independently?

f. How would car choices change if the two acted "jointly"? Would they get faster or slower cars than when they act individually?

g. How would car choices change if the two acted "sequentially," so that one individual bought his car first and the second bought his car with knowledge of the first individual's choice? Who would buy the faster car?

h. If the conditions in both 2c and 2d are true, should the government intervene and restrict the "arms-race" as a means of improving welfare? Under what conditions would stopping the "arms-race" be optimal?

3) *True, False, or Uncertain*: Making it easier to quit smoking (say by the introduction of a new pill that reduces the negative effects of quitting) can increase the number of smokers and the total quantity of cigarettes smoked.

4) *True, False, or Uncertain*: Rice is an inferior good.

5) *True, False, or Uncertain*: A $100 weekly tax on full-time jobs (30+ hours per week) would cause workers to work fewer hours per week.

6) *True, False, or Uncertain*: The availability of ebooks reduces the sales of physical books.

7) *True, False, or Uncertain*: The benefit to a consumer of a price cut depends only on the amount he or she consumes of the good, and not the willingness to substitute from other goods.

Part II

Market Equilibrium

Chapter 7

Discrete Choice and Product Quality

Imagine a setting in which individuals are deciding whether to buy either 0 or 1 of a good. Consider Figure 7-1.

Here we ignore that a lot of goods that look discrete can be thought about as more continuous. Multiple haircuts are purchased over time, for example.

MARKET DEMAND IS A DISTRIBUTION FUNCTION

Each person has a cutoff value v_i, which gives the value they place on the good in dollars. At v_i, individual i is indifferent about buying the good. This can be thought about empirically as well. We can infer v_i, for instance, by looking for the price at which a person moves from not buying to buying.

We can further consider a distribution $F(v) = \Pr(v_i \leq v)$, which gives the fraction of the population with a value less than v. Then the market demand at a price P will be given by $D(P) = (1 - F(P))N$, where N is the number of people in the population and $1 - F(P)$ is the fraction of people who buy.

Suppose we consider a normal distribution of individual values. What would the demand curve look like? See Figure 7-2. The function asymptotes toward N, which is the maximum number of units that could be sold. (People with a negative value do not want the good at $P = 0$, so demand may be less than N for any positive price.)

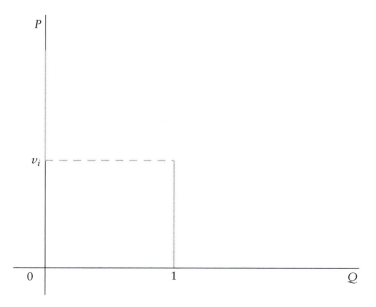

Figure 7-1: At price v_i, one is indifferent between buying the good or not.

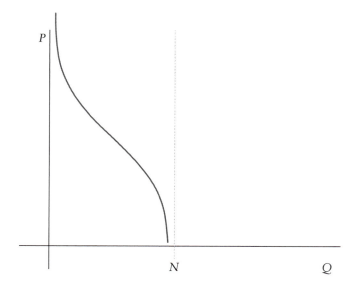

Figure 7-2: Normal distribution case.

What if the distribution is uniform? That is, $V \sim U(A, B)$. At B, no one buys, and at A, everyone buys. The demand curve is depicted in Figure 7-3. Note that it is linear. Linear demand curves can often be thought of as representing a uniform distribution of preferences.

The elasticity of demand for a general distribution is given by

$$\frac{P}{D(P)}\frac{\partial D(P)}{\partial P}=-\frac{PNf(P)}{(1-F(P))N}=-\frac{Pf(P)}{1-F(P)}$$

where $f(P)=F'(P)$ is the probability density function corresponding to F.

The elasticity of demand depends on the fraction of people on the margin versus the fraction of people already buying. At a point with high density, demand will be relatively elastic. We can think about these demand functions as adoption curves; that is, as new goods come out, these curves tell us how many people will buy as the price decreases over time. Suppose appliances are getting cheaper; consider the demand curve for the normal distribution in Figure 7-2. Once the price gets low enough, demand is very elastic, so additional reductions in price cause large changes in purchasing. The fact that the middle class goes on a spending spree when appliances get cheap enough isn't necessarily a bandwagon or network effect. For cell phones, this model might not be enough. Cell phones might additionally require a network effect explanation because more people owning cell phones raises non-owners' desires to own cell phones.

Now note that whenever we have a demand curve, we can approximate it using a linear demand curve. This type of procedure is not always sufficient for the analysis we might want to do—in monopoly problems, for instance, often the curvature of the demand curve can be important. For many purposes, however, if we're interested in the price-quantity relationship, honing in on a small area will be useful. See Figure 7-4. The question is about whether we're approximating all the elements relevant for behavior. For assessing welfare, for instance, we can look at the budget constraint as an approximation to an indifference curve. If we're analyzing substitution, however, a linear approximation of the indifference curve means we get perfect substitutes. So for this, we'd have to model the curvature of the indifference curve.

EQUILIBRIUM PRODUCT QUALITY

Once we're operating in the world of single choice, we can solve more complicated problems. Now we will stay focused on the discrete choice but extend the model to think about varying levels of quality. Let

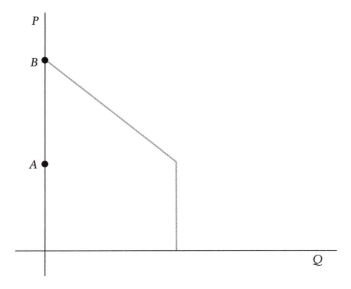

Figure 7-3: Uniform distribution case.

q = quality. People will still just buy one unit of the good, such as a TV, but now they will also consider what degree of quality they want in their TV. For simplicity, we will assume quality is continuous. For the TVs, if we think about size, this means that you can buy a 59.1-inch TV. This isn't particularly realistic, but it will make our lives easier.

We'll consider indifference curves $U(X, q)$, where X denotes other goods, $P_X = 1$, and M = Income. So $X = M - P$, where P is the price of the TV. This means we can rewrite the utility function

$$\bar{U}(M - P, q).$$

Sometimes we assume quasilinear utility $U(X, q) = X + V(q) = M - P + V(q)$. This means, when M is high enough that X is consumed, there is a constant marginal utility of income, since utility is linear in X.

Pick q to max $V(q) - P(q)$, where $P(q)$ is the schedule showing the price for each quality level. This means quality is not a normal good under quasilinear utility—M doesn't matter for the choice of q. Though quasilinear is a popular model, we know quality choice increases with income. We can use a more general utility function, $\bar{U}(M - P, q)$, to get quality to be a normal good.

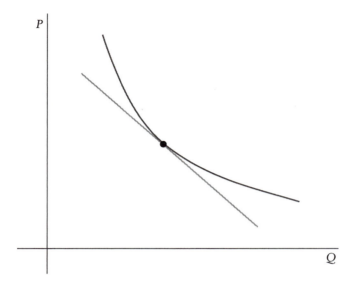

Figure 7-4: Approximating the demand curve with a linear demand curve.

Why do people increase quality (rather than quantity) as income goes up? There are physical constraints—stomach capacity, hours in the day, and so forth. How might we model this quantity-quality problem? Consider

$$U(X, NV(q)) + \lambda[M - X - NP(q)]$$

where N denotes quantity. We can rewrite this as

$$U(X, Z) + \lambda \left[M - X - Z \frac{P(q)}{V(q)} \right]$$

where $Z = NV(q)$ denotes "effective consumption," since it summarizes both quantity and quality. There's a q^* that marks the efficient quality level (the ratio $\dfrac{P(q)}{V(q)}$ denotes cost per unit enjoyment). The world puts some limits on these parameters. For example, we may have $N \leq \bar{N}$. $\bar{N}$ could be the capacity of your stomach, for instance. Then as income rises, people will increase quality. This might explain why poor people don't just buy fewer high-quality goods; instead, they save money by purchasing goods of lesser quality.

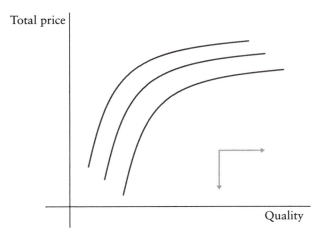

Figure 7-5: Consumers prefer higher quality and lower price.

Now consider a slightly more general model, where q_j = quality of j and N_j = consumption of j. We can write the utility function

$$U\left(\sum_{j=1}^{J} N_j, \sum_{j=1}^{J} N_j q_j, X\right).$$

In this utility function, an individual cares about total amount and quality-weighted amount. An interesting issue is whether $\sum_{j=1}^{J} N_j$ has positive or negative utility. Implicitly, it has negative marginal utility, because it's a constraint (limited by $\bar{N}$). Maybe the goods are various types of foods, and quantity is measured by the calories they deliver, and quality is measured by their taste or eating experience. People enjoy eating but do not want too many calories, at least if their income is high enough that starvation is not a concern. In general, so long as demand for $\sum_{j=1}^{J} N_j$ grows slower than $\sum_{j=1}^{J} N_j q_j$, you'll move up the quality ladder as you consume more.

In general, people prefer lower prices and higher quality. See Figure 7-5. There is no reason the curves here need to be concave, but let us suppose they are. How will we describe consumer behavior in this model?

When the consumer goes to the store to buy a TV set, we will consider a price $P(q)$, a price increasing in quality q. The consumer will choose a point where his or her indifference curve is tangent to this price line from below. See Figure 7-6.

The indifference curve always has to be concave relative to the equilibrium price line, which is why it did not matter that we assumed

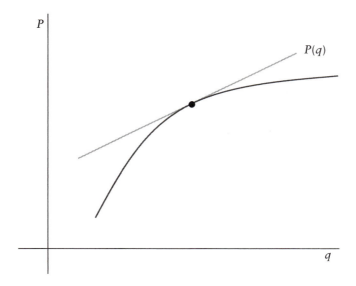

P

P(q)

q

Figure 7-6: Slope gives us the marginal willingness to pay for an additional unit of quality.

indifference curves were concave. (If part of the indifference curve is more convex than the price line, the consumer will never choose that point in equilibrium.) The slope near the tangency will give us the marginal willingness to pay for an additional unit of quality.

Now let's think about the firm side of the market. Assume there are a large number of producers and that the unit cost of production is $C(q)$. Each producer makes one unit and chooses the quality to produce. Let N = number of consumers and M = number of producers. Assume $M > N$. This implies profits $\Pi = 0$. Why? Some producers are not going to produce in equilibrium, which means they earn 0 profit. This means the producers who are producing must be making 0 profits. Thus $P(q) = C(q)$. Now, as before, consumers pick a level of quality where their indifference curves are tangent to the price curve. See Figure 7-7.

What happens if someone gets richer? Their indifference curves will get steeper, because they will have a higher preference for quality, and quality is a normal good. They will cross the indifference curves of the poorer consumers from below. See Figure 7-8.

Heterogeneity among consumers pushes them to consume at different points along the producer's cost curve. The price differentials at each point represent each consumer's marginal willingness to pay for quality and also trace out the producer's overall marginal cost of quality.

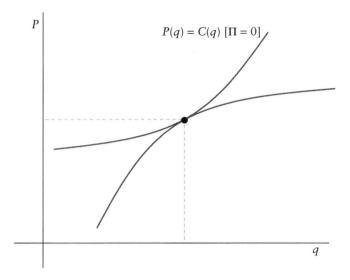

Figure 7-7: Consumers pick a level of quality where their indifference curves are tangent to the price curve.

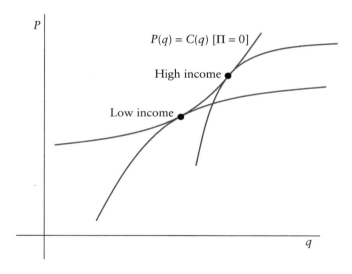

Figure 7-8: Consumers pick higher levels of quality as they become richer, since q is normal.

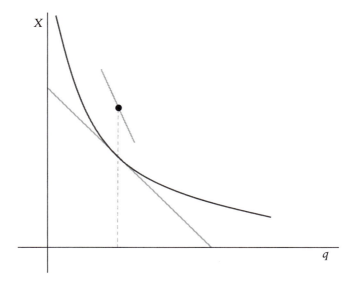

Figure 7-9: If q is normal, then the willingness to pay for q rises as X increases, holding q fixed.

Let's think a little more about what q being normal implies about indifference curves between other goods X and quality q, holding N constant. Consider Figure 7-9, and pick a point on the indifference curve. Then as we increase vertically from this point, it must be that the indifference curves become steeper, because this is what would drive additional consumption of quality. That is, q normal implies that willingness to pay for q rises as X increases, holding q fixed. The normality condition for X would be the reverse. As one moves right along a horizontal line from the chosen point along the indifference curve, the indifference curves must be getting flatter, because this would drive increased consumption of X. Note that this is true in the two-good case but more complicated in the multi-good case.

Thus, normality of q means that

$$\frac{\partial(U_q/U_X)}{\partial X} > 0.$$

HETEROGENEOUS FIRMS

Consider again the quality indifference curves for two consumers, depicted in Figure 7-10.

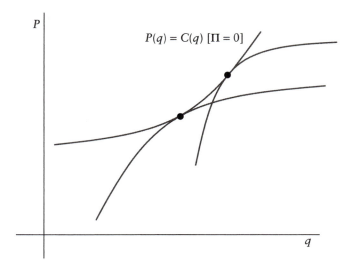

Figure 7-10: Different allocations along the quality cost curve.

Different consumers sorted out along the cost curve. This was the case of two different consumer types and one firm type. Now what if we change the problem to consider two different types of firms and one type of consumer? Assume free entry of both types of firms, so there are many firms of type 1 and many firms of type 2. Would we see more than one point in equilibrium? See Figure 7-11. Unless it is a knife-edge situation, where the price of additional quality consistent with zero profits exactly coincides with consumer willingness to pay for quality, we would end up with a single point. In other words, one of the types of firms offers a better deal; the other type cannot provide consumers with the same utility without taking a loss. Profits of the former firms would be driven to 0, and consumers would buy from the firm that better satisfied their needs. Free entry of firms guarantees this.

What happens if there is a limited supply of firms? Suppose $N_1 + N_2 > N_{cons}$, $N_1 < N_{cons}$, and $N_2 < N_{cons}$. So, there are fewer firms of each type than total consumers, but both firms together can produce enough to satisfy all the consumers (we're still in the discrete production case). There will be a unique equilibrium. People prefer to consume from firm 2, but firm 2 cannot satisfy everyone. So all firms of type 2 will produce, but some type 1 firms will be required. But since $N_1 + N_2 > N_{cons}$, type-1 firms must earn 0 profits. No firm of type 1 can make profits in equilibrium, otherwise other firms would enter. Firms

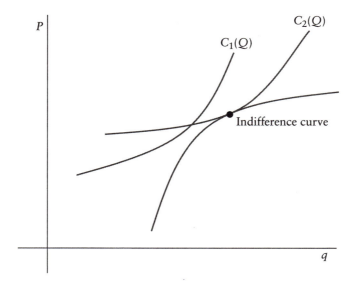

Figure 7-11: Ignoring capacity constraints (for the moment), homogeneous consumers buy from firms of type 2.

of type 2 are constrained to offer a combination that's at least as good as the offerings by the type-1 firms. So, they produce along the curve where they are also tangent to the consumers' indifference curves.

In other words, we draw the equilibrium by first tracing out the combinations of price and quality that yield zero profit for type 1, which is the cost curve $C_1(q)$. See Figure 7-12. Then we draw the indifference curve tangent to it, because any type-1 firm deviating from the tangency point must either take a loss or get no customers (who prefer to take the tangency point offered by other type-1 firms).

Finally, the type-2 firms maximize profits subject to the constraint that consumers are no better off buying from a type-1 firm instead. Note that the $\Pi_2 = \Pi^*$ curve is not its cost curve, but is above it because they are able to charge more than cost.

Consider more generally what the firm's indifference curves look like, depicted in Figure 7-13. The positive profit curve in Figure 7-12 is just the cost curve shifted upwards.

Is it the case that the higher quality producers must be the ones making a profit? No. The reason that this was true came from the way we drew the curves. Consider the equilibrium that would result from the situation depicted in Figure 7-14.

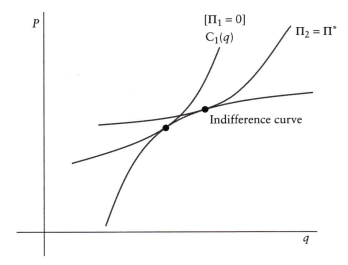

Figure 7-12: Firms of type 1 make no profit in equilibrium. Firms of type 2 produce according to $\Pi_2 = \Pi^*$ to maximize profits.

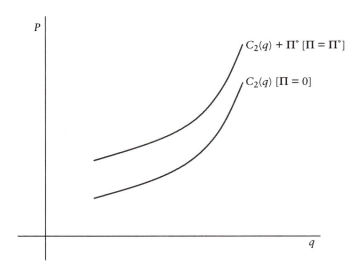

Figure 7-13: Indifference curves are just vertically shifted and thus have the same slope at every point.

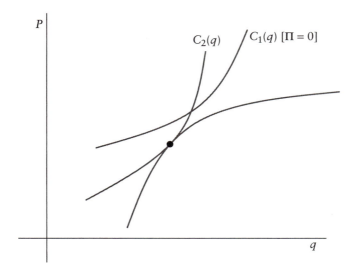

Figure 7-14: In equilibrium, the firm earning positive profits produces a lower quality good.

HETEROGENEOUS FIRMS AND CONSUMERS

Now, broaden the question to multiple producer types and multiple consumer types. We will continue with the discrete case. Think about consumers A, B and firms 1, 2. B prefers high quality more than A (i.e., B's indifference curves are steeper). Then assume $C_2(q) < C_1(q)$ and $C_2'(q) < C_1'(q)$. That is, cost curves of 2 are below and flatter than those of 1. Finally, assume $N_1 + N_2 > N_A + N_B$ (i.e., the number of producers is greater than the number of consumers), $N_1 < N_A + N_B$, and $N_2 < N_A + N_B$.

This tells us a lot about the equilibrium right away. We know $\Pi_1 = 0$, $\Pi_2 > 0$, because type-2 producers have lower cost curves. Further, type 2 produces higher quality because their marginal cost is lower, and B's want to buy from 2. Who buys from type 1? Some A's, at least.

Only one type-1 isoprofit curve is relevant for drawing the equilibrium ($\Pi_1 = 0$), so we begin by drawing that one. The type-A indifference curve must be tangent to that isoprofit curve at the equilibrium quality purchased by those consumers (Q_{1A}), otherwise that quality would not be profit maximizing. See the lower left point in Figure 7-15.

Next, we check whether any single type of firm sells to both types of consumers. Reaching a conclusion here requires an additional

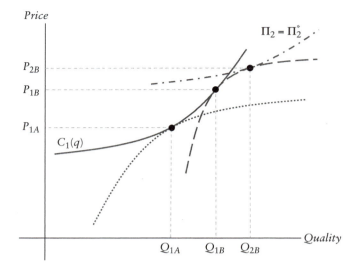

Figure 7-15: We've assumed there are more type-B consumers than there are type-2 firms. So, all type-B consumers buy from type-2 firms (the topmost point), but then remaining type-B consumers buy from the type-1 firms (the middle point). Type-A consumers buy from type-1 firms.

assumption, which we take to be $N_B > N_2$. That is, there are more B consumers than the type-2 firms can supply. Then only type-B consumers buy from firm 2 because they have the stronger preference for quality. With some type-B consumers also buying from firm 1, equilibrium requires that the two transactions be on the same type-B indifference curve, which is tangent to both the type-1 isoprofit curve and the type-2 isoprofit curve at Q_{1B} and Q_{2B}, respectively. This is the equilibrium depicted in Figure 7-15.

What if we change this last assumption, and instead we have $N_B < N_2$? Now all B consumers buy from firm 2, but some A consumers will also buy from them. This equilibrium therefore has less profit than shown in Figure 7-15 (i.e., the equilibrium isoprofit curve is below what is shown in the figure).

What if $N_1 + N_2 < N_A + N_B$? Now, some consumers do not get served. Boundary conditions will be determined by the utility side rather than the cost side. We'll now need a baseline utility figure to determine what level of utility is received by consumers who do not buy anything.

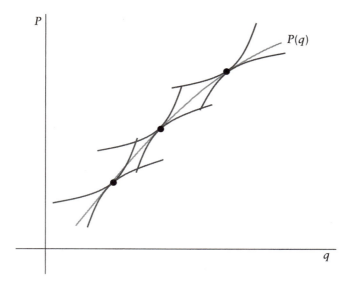

Figure 7-16: Producers are tangent to consumers, and we can trace out a function $P(q)$ from the tangencies.

Note that when multiple consumers buy from the same producer, we follow the producer's cost curve, and we follow the consumer's indifference curve when the same consumer must be indifferent about buying from different producers. With a continuum of types on both sides, you get a more general picture, depicted in Figure 7-16.

Chapter 8

Location Choice

*An Introduction to Equilibrium
Compensating Differences*

Now, think about a model of travel time to the center of a city. Let t be the travel time for living at t, and $R(t)$ be the rent for living at t. In equilibrium $R'(t) < 0$; that is, rents become cheaper as distance from the city center increases because everyone prefers less travel time. We will not want to put travel time in the utility function, however. People do not directly care about how far they have to travel; they care about how much travel will detract from leisure time. We'll consider a simple utility function $U(C, L)$ and budget constraint $C = (24 - L - t)w - R(t)$, where L is leisure time and C is consumption. We can write the Lagrangian

$$\mathcal{L} = U(C, L) + \lambda[(24 - L - t)w - R(t) - C].$$

This yields the first-order conditions $\dfrac{\partial U}{\partial C} = \lambda$, $\dfrac{\partial U}{\partial L} = \lambda w$, and $-R'(t) = w$.

So, the optimal choice of where to live has savings in rent from living another hour farther away that is equal to the wage rate. Thus, we see that we've made the restriction that travel is really the same thing as work: should one work by driving one's car, or should one work at the workplace? It seems odd that being in the car would produce anything, but it does through the market: living farther allows someone else to live closer to the workplace and potentially spend more time there.

Now we'll want to consider what the rent curve looks like in this model. There will be a level of rent paid by people at the center of the city $R(0)$. The slope at that point will be the wage of the highest wage

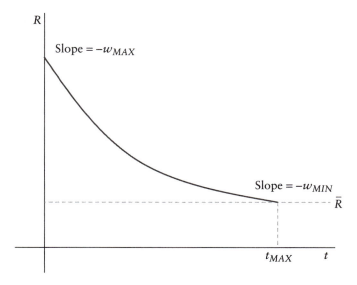

Figure 8-1: The rent gradient. $\bar{R}$ denotes what can be earned from using the land for agriculture, for example, rather than housing. In other words, $R(t_{MAX}) = \bar{R}$.

worker, $-w_{MAX}$. Thus, the curve is not only downward sloping; it will be convex, because people's wages decrease as we move down the curve.

Holding L constant, we can also draw a person's indifference curves in Figure 8-1 by using part of the budget constraint: $(24 - L - t)w - R$. Those consumption-constant curves would be straight lines with slope $-w$. A w person's optimal location choice is the t where the indifference curve is tangent to the rent gradient shown in the figure.

Now, how do we solve for $R(0)$, the initial boundary condition? We know the slope at that point but not yet the level. Let the boundary of the city be the distance from city center needed to fit all of the city's residents and let t_{MAX} be the amount of travel time required to reach the city boundary. In other words, at t_{MAX} we run out of people who are driving to the city center. At that point, there is a lowest rent that will be set, for instance, by what can be earned by using the land for agriculture. Call this rent $\bar{R}$. Thus, $R(t_{MAX}) = \bar{R}$, and then we know all the slopes. So, we can follow our first-order conditions about rent toward the city center until we reach the level of rents in the center and find $R(0)$.

Note that the function of equilibrium rental prices must be continuous at $\bar{R}$. If there were a jump between $R(t_{MAX})$ and $\bar{R}$, the person living at t_{MAX} could live a nanosecond farther from work and receive a discretely lower rental price. This could never be an equilibrium.

PROPERTIES OF THE RENT GRADIENT MODEL

The rent gradient model is illustrative of the fact that we can get a lot out of a model where preferences are very simple and we conceptualize a consumer choice problem as a production problem (in this case, cost minimization). Determining where to live had nothing to do with preferences; we simply compared the wage cost of traveling farther with the rent saving. We could do the same thing for thinking about which car to buy, where the agent trades off between space and gas mileage. Moreover, the rent gradient model illustrates how hedonic models work. That is, we had that consumers preferred less travel time and less rent to more, implying that the equilibrium is going to have to be downward sloping. An individual must be compensated with lower rent for needing to travel farther, and vice versa. So everyone will choose a point where they are indifferent about moving a little bit closer. Then the rent gradient is convex, because the highest-wage consumers live close to the city center and the lowest-wage consumers live farther.

But now we just have a downward-sloping, convex curve. How did we pin down the exact curve? We looked at the boundary condition. We considered a world in which there is some lowest rent value $\bar{R}$ set by what the land could earn for some use other than housing. Let's consider some comparative statics. Suppose we raise the wages of the lowest-income people. The slope of the rent gradient would be steeper for the lowest-income people, but the slope would not be affected elsewhere. Thus, rents would be higher throughout the distribution. This result is depicted in Figure 8-2.

What would happen if we raised incomes of just the top half of the distribution? There would be no effect on the lower half of the income distribution, but the rent gradient for the upper half would get steeper, and rents would rise.

What would happen if we raised inequality, holding the average wage fixed? Since the average wage is fixed, the new rent gradient has the same end points. Note that a given person's rent is determined by the wages of the people who live farther from the city center (and earn less) than he or she does. So, because those people's wages have gone down on

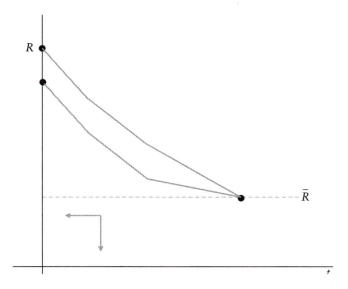

Figure 8-2: Raising wages for low-income workers increases the slope of the rent gradient for wages near $\bar{R}$.

average (while wages have increased for people living closer and making more), it must be that his or her rent is lower. See Figure 8-3.

Note also that the average is weighted by distance. If roughly the same number of people live at each distance away from the city center, we get the normal average. But if we think about each fixed distance from the center of the city as a circle, we might get more people living at a given distance as we move away from the center.

Consider another practical application where the locations differ in terms of their crime rate, or some other housing quality metric other than distance from the city center. Let's assume poor people live in high-crime areas. Let's assume we focus on people not committing crime. Why would they live in high-crime areas? Low-crime areas are normal goods. Now, what if we came in and reduced crime in those areas? The people could be worse off. The rent they pay, determined by the marginal person willing to live in that neighborhood, might very well rise. And a given individual might value the reduction in crime less than the marginal people. What about a policy where we increase the quality of housing? Again, we might easily get that people are unwilling to pay the higher rents and simply move elsewhere, back into lower-quality housing.

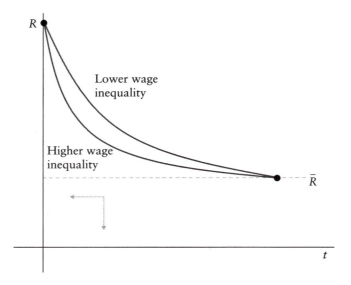

Figure 8-3: Raising wage inequality lowers rents almost everywhere (not the endpoints).

Note that the reservation rent pinned down a lot for us in this model. We could alternatively consider a world where supply of high-quality housing is very elastic. Consider Figure 8-4. The elastic supply of high-quality housing pins down $\bar{P}$, and prices for lower-quality housing will be determined by the integral over the marginal willingness to pay of buyers for lower-quality housing.

As discussed previously, each person has indifference curves that can be drawn (downward sloping) in the figure. A person's optimal location—this time in terms of a crime rate—is where his or her indifference curve is tangent to the price gradient. For the poor person, this is probably at a point in a relatively high-crime neighborhood: to the right in Figure 8-4.[1]

Consider the policies we considered before, where the lowest-quality locations are enhanced to be the same as, say, at point A. Then those locations now have to rent for the same as point A, so policy has cut off the rent-versus-location opportunities as in Figure 8-5. The poor person may not locate at A, but has to have utility as if located at A because the locations to the right of A are, by regulation, no different. But he or she gets less utility at A than when the neighborhood B had more crime because that extra crime allowed rents to be lower than at A.

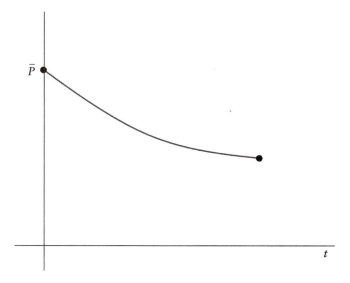

Figure 8-4: Very elastic supply of high-quality housing pins down $\bar{P}$ (quality is decreasing with location along the x-axis). The price paid by someone farther out will be the integral over the marginal willingness to pay of the higher quality people to his left.

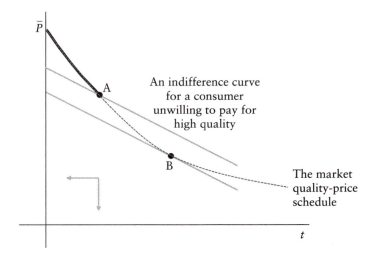

Figure 8-5: A minimum-quality regulation restricts consumer choice. The more convex dashed curve is the market quality-price schedule (quality is decreasing with t), much of which is prohibited by the regulation.

Lower-income buyers typically do not buy low-quality things because they like them. They buy them because they dislike them less than do higher-income people. These kinds of models help us see why that is.

Let's consider this housing model from the perspective of the model for quality that we covered earlier in the book. Suppose there are two qualities of neighborhoods, high and low. Suppose there are 60 high-quality spots and 40 low-quality spots for 100 total consumers. If we shift supply so that there are 70 high-quality spots and 30 low-quality spots, while assuming the price for high quality is pinned down (due to high elasticity of supply), then all the low-quality consumers will be worse off. The reasoning is that rather than the person with the 60th highest willingness to pay being indifferent between high- and low-quality housing, now the person with the 70th highest willingness to pay will be indifferent between high- and low-quality housing, so the price of low-quality housing must increase. To put it mathematically, initially we have $P_H - P_L = V(60)$. When we increase the number of spots in H to 70, Then $P_H - P_L = V(70)$. But P_H stays the same; then since $V(70) < V(60)$, P_L increases. People living in the low-quality neighborhood are worse off.

Can we use the rent gradient model to think about urban density? Yes, if there is an increasing marginal cost for higher buildings. Your city would look like a city—that is, homes would get taller as you got closer to the city. Buildings will be of a height such that the marginal cost of an additional floor equals the value of a home at that location—which is higher closer to the city. We can also complicate the model by adding H for house size; that is, how much house you buy. Then an individual's first-order condition becomes $w = -R'(t)H$. Whether high-income people live near the city center or farther away depends on whether the elasticity of demand for H with respect to income is greater than or less than 1. If that elasticity is greater than 1, people with high income want bigger houses, so they live farther from the city, where land is cheaper.

Chapter 9

Learning by Doing and On-the-Job Investment

In school, one learns general principles. On the job, one has to take these general principles and focus on something more specific. Someone who has just finished medical school learns how to apply these principles to patients in their residency. An electrical engineer might decide to specialize in car repairs. This is what we will call on-the-job training or investment. There are two models for this: learning-by-doing and an explicit investment model. Often, we will have a data problem. Suppose we asked how much on-the-job training there was in the United States. We might look at companies. But companies frequently do not publish information about how much they are investing in their workers. Companies are not required to report this, and instead they can deduct it as labor expense. They keep accounts of capital that they must depreciate over time, but the same is not done for human capital. Often companies do not even have internal information on this.

HUMAN CAPITAL ACQUIRED FROM TRAINING PROGRAMS ADMINISTERED BY THE EMPLOYER

If workers are paying for their training, however, how would they do it? One way would be that they are receiving lower earnings, net of their training costs. Consider Figure 9-1 below. This is the explicit investment model.

Evaluating the human capital investment amounts to comparing the two areas in the chart: whether the earnings gains later in life justify the

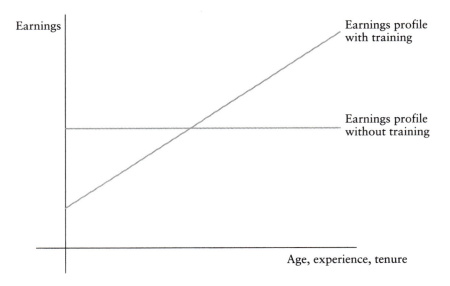

Figure 9-1: People who receive on-the-job training expect to earn lower wages today because they spend part of their time enrolled in training, and higher wages tomorrow when their time is more productive due to the training received.

costs incurred early on. Alternatively, if labor-market data were showing us these two curves, we could infer how much human-capital investment people were doing. The investment amount is related to the left-hand area, although it is not identical to it.

LEARNING BY DOING

The other model is learning by doing. Learning happens as an apparently automatic byproduct of working. There is no tuition bill, and the learning does not require any time off from production or any deliberate slowdown from production in order to learn. It is tempting to conclude from this observation that the learning is free. In other words, the wage profile might look like the upper profile in Figure 9-2 below, where the training never involves having a wage below the alternative.

This looks like a free lunch. But we know in economics there is no such thing as a free lunch. Sometimes the cost of the lunch is hidden. So where does the cost enter in here? The market equilibrium eliminates the free lunch. If the wage profile were the upper one in the figure, then everyone would want it, and no one would want the horizontal profile for the jobs without learning. The equilibrium profile for the

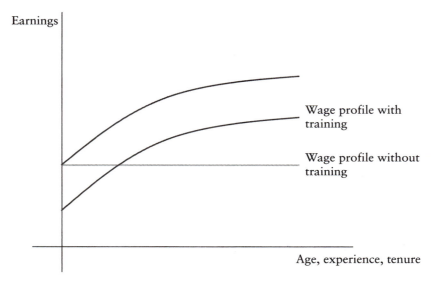

Figure 9-2: Learning by doing is not free because workers compete to get it. The wage profile (and therefore earnings profile) ends up being below the no-training alternative for a period of time, much like it is in the investment model.

job with learning has to be lower (see the lower profile in the figure), or the horizontal profile higher, in order for people to be willing to do both jobs. So, the final learning-by-doing picture (Figure 9-2) ends up looking a lot like the investment picture (Figure 9-1). Indeed, Sherwin Rosen wrote a paper arguing that the two pictures are identical (Rosen 1972). This answers one of the questions posed in the introduction to this book.

Notice that Figure 9-1 does not show wage rates, which reflect earnings per hour worked, whereas Figure 9-2 does. The training case has the added complexity as to what we mean by an hour worked: does it include time spent in the training program? If not, then the starting wage rate might not be particularly low in comparison to the starting wage rate to be earned without training. But starting earnings would be, because the training takes up time.[1] With learning by doing, there is no time spent training so this distinction does not come up. Both wage rates and earnings start out low in the learning-by-doing scenario because workers compete to be in a position that automatically gives them skill.[2]

TYPES OF HUMAN CAPITAL

We can think about different degrees of specialization. There is firm-specific investment—investment raises productivity in one firm but not in other firms. Then there is industry-specific investment; these skills would be equally useful at any firm in a given industry. In the industry-specific case, who pays for the training? One might be tempted to say there is a positive externality because a result of a firm's investment in its worker's industry-specific skills may be that the worker applies the skills at another firm in the same industry. However, because both the firm and worker recognize this, the worker's initial wages will be lower, and the worker pays for the training.

What would be the evidence for industry-specific investment at a particular firm? If that firm goes out of business and their employees go to firms in the same industry, as Derek Neal found in his study of displaced workers (Neal 1995), then at least part of the training that was acquired was likely applicable elsewhere in the industry. How do we tell whether there was firm-specific investment? We would look at wages. If the firm goes out of business, and the employees must accept reduced wages elsewhere, then part of their human capital may have been firm-specific. Note that if the industry is small, and the company that fails is sufficiently large, there might be supply-side effects that would need to be controlled for.

Now suppose we have a monopoly, and we are thinking about industry-specific investment. The monopoly might be willing to incur these costs because it is the only firm in the industry. Note that even here, however, the worker may still pay some of the costs. To see this, suppose instead that the monopolist paid for the entire investment, and now the worker is trained and productive at the firm. The worker threatens to leave unless given a raise. It is a largely empty threat, because those skills are not useful elsewhere. But still there may be some bargaining and the trained worker ends up getting paid more. But then the untrained workers anticipate getting paid more after the training and compete to have the position—as in the learning-by-doing case, the equilibrium result is lower earnings during the training phase.

Now we can think about firm-specific investment more generally. How does the worker know whether firm-specific investment will yield higher wages at the company later on? If the firm has done this for its skilled workers in the past, this would be a helpful signal. But in general we expect bargaining between the firm and the worker over this.

We can also look at turnover. Turnover of workers is very heavy early on and much less later on. This is similar for industry turnover. With firm-specific investment, we often think it is efficient that the workers skilled at a firm stay at that firm; turnover should be low. Separation between the employer and the worker occurs when the relationship is no longer efficient.[3]

The firm- and industry-specific distinctions are also useful for thinking about other factor markets, such as markets for raw materials, or intermediate inputs. If a user of the factors goes out of business and the factor suppliers switch to supplying someone else in the industry, then these suppliers had acquired some amount of industry-specific capital.

Chapter 10

Production, Profits, and Factor Demand

COMPARATIVE ADVANTAGE AND THE PRODUCTION-POSSIBILITY FRONTIER

Now we turn to a treatment of production. We will consider Robinson Crusoe, who lives on an island and has two plots of land. On Plot A, Crusoe can produce 10 bananas or 5 oranges (or some mixture of the two). On Plot B, he can produce 15 bananas or 40 oranges (or some mixture of the two). What is Crusoe's production-possibility frontier? See Figure 10-1.

On Plot A, an orange costs 2 bananas. A banana costs ½ of an orange. On Plot B, an orange costs 3/8 of a banana. A banana costs 8/3 of an orange. Plot A is the low-cost producer of bananas, and plot B is the low-cost producer of oranges. This is what an economist calls comparative advantage. We don't care how good in absolute terms the plots are at producing bananas and oranges. A producer could be the low-cost producer because it is really good at producing a particular item or really bad at producing other items.

So, we get a convex production-possibility frontier. The marginal cost of producing oranges rises as we produce more oranges. Why is it rising in this model? Two features are important. First, there is heterogeneity in plots, and second, we use the lowest-cost methods first. This is why we frequently assume increasing marginal cost of production. As we use more and more resources for production, we are forced to use resources that have less comparative advantage.

106

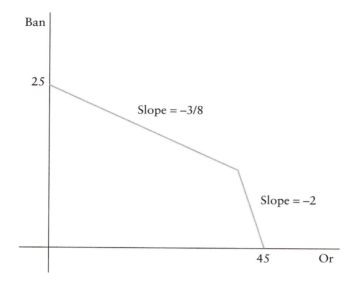

Figure 10-1: Production-possibility frontier. Different slopes reflect cost differences across plots.

Note that Plot B can produce more bananas than Plot A. Nevertheless, it is a waste for Plot B to produce *any* bananas unless people want more bananas than Plot A can possibly produce.

For the same reason, we're not talking about the price of the plots. The price will be based in part on absolute productivity. Really bad land might be cheap to buy, but it can still have a comparative advantage over expensive plots of land.

With three plots of land, note that we get another kink. As we get many plots of land, we will get a smoother convex shape, as in Figure 10-2. And the more heterogeneous the plots of land, the more convex the shape. If Robinson Crusoe is by himself, he will find the tangency of the production-possibility frontier with his indifference curve. The slope of the tangency will give the equilibrium price.

Now let's suppose Crusoe is allowed to join NAFTA (i.e., an agreement to trade in a world market). They will tell Crusoe the price of oranges in terms of bananas (or the reverse) in the world market. He wants this price to be *very* different from the price set from his tangency. If he gets the same price as his current price, he is no better off. Let's suppose that NAFTA tells Crusoe that oranges are more expensive in the world than on his island. This sets a trade price line with slope $-P_{OR}^{NAFTA}$,

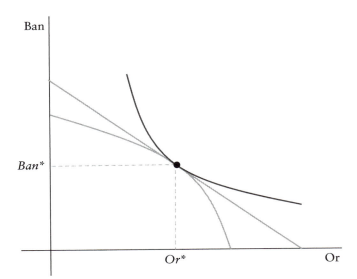

Figure 10-2: Crusoe will produce where the production possibility frontier is tangent to his indifference curve. Marginal cost will be equal to marginal value.

depicted in Figure 10-3. In this setting, Crusoe will produce more oranges and less bananas $(OR^*_{PROD}, BAN^*_{PROD})$ and then trade to get $(OR^*_{CONS}, BAN^*_{CONS})$, which is on a higher indifference curve. (Similarly, if NAFTA says bananas are more expensive, Crusoe will produce more bananas and trade a bunch of them for oranges; he will still be better off.)

So, what do we get out of this model?

1. Trade is good. Robinson Crusoe is better off when part of NAFTA.
2. Price difference implies gains from trade. The more Robinson Crusoe's price differs from NAFTA's price, the better off he will be.
3. A theory of the firm. The firm owns a production process and chooses $(OR^*_{PROD}, BAN^*_{PROD})$. What allows this to be a theory is that we know the firm will pick $(OR^*_{PROD}, BAN^*_{PROD})$ regardless of the owner's personal preferences.
4. A competitive firm produces where price equals marginal cost. Note that this becomes an inequality in corner cases. If the price line is steeper than the production-possibility frontier for any tangent line along the frontier, then Crusoe will produce only oranges; and further, we will not have that marginal cost equals marginal value.
5. Marginal cost = marginal value. Again, this may not hold in corner cases.

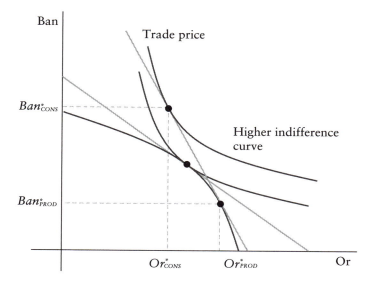

Figure 10-3: A large difference between the trade price and the price Crusoe would face without trade makes Crusoe much better off with trade.

This production example has a lot in common with chapter 7's result that the poor could be harmed when their neighborhood gets safer. In both examples, the benefit from trade comes from being able to make different choices from everyone else. Crusoe produces oranges and trades them for the bananas (produced by others) that he wants. A poor person may take the low-quality housing in order to get more of the other goods that he wants, while in the same market richer people give up other goods in order to get the high-quality housing that they want. Trade makes these choices possible.

THE PRODUCTION FUNCTION

Now we will want to consider how a firm behaves. Consider the following production problem. We will have a production function, output Y, and inputs $X_1, \ldots, X_N$, and we will specify

$$Y = F(X_1, \ldots, X_N).$$

The production function is the result of maximization already—F returns the maximum amount of output achievable using the inputs. Just going out and purchasing the inputs by itself will not make the output magically appear. The inputs need to be put together in the right way. The details of managing and combining the inputs can be interesting, but whenever we start with a production function we have put those details aside. Nevertheless, you will see that we have some interesting things to say about how firms interact with the rest of the marketplace.[1]

The production function is said to exhibit constant returns to scale if the function is homogeneous of degree 1 in the inputs. Equivalently, if we calculated the elasticity of output with respect to each input, those elasticities would sum to 1. If the sum is less (greater) than 1, then the production function is said to exhibit decreasing (increasing) returns to scale, respectively. Decreasing returns to scale is a common assumption to make when looking at a particular firm.

We will also assume the firm is competitive. That is, the firm does not affect (1) the price of output P or (2) the prices of inputs, $w_1, \ldots, w_N$. Is there a difference already between this and utility? We can measure the output here, whereas we couldn't do this with utility.

PROFIT MAXIMIZATION

Firms will maximize profits. Why is this a reasonable assumption? In consumer theory we say that more income permits an individual to obtain a higher utility level. An individual owning a firm therefore gets more utility the more profit income that he or she obtains from that firm. If, on the other hand, the activities of the firm directly enter the person's utility function—having lots of capital, for example—this will not be a good assumption.[2] Owners of basketball teams typically care about winning; it is not necessarily a priority for them to have the most amount of money possible at the end of the season. Gary Becker's dissertation was about the idea that some firms like to hire some people more than others—for example, they might want to discriminate (Becker 1957). So the profit-maximization assumption may miss some of what's going on, but it is typically a good first-order approximation to think about firms as maximizing profits. We will denote profits as revenue minus cost, $PY - \sum_{i=1}^{N} X_i w_i$, and maximum profits as

$$\max_{X_1, \ldots, X_N} PF(X_1, \ldots, X_N) - \sum_{i=1}^{N} X_i w_i.$$

This looks a lot like our utility maximization problem. But here you can trade the output. You can't trade utility. Even if we could quantify utility, these problems would still be different. Production of utility is intrinsic to consumption. Output, on the other hand, is ordinal, measurable, and transferable. This is why we achieved so much more traction with the rent gradient model after turning it into a production problem. In that model, people ended up working (spending their time) where it was most productive—either in their car or at work.

The profit-maximization problem leads to the first-order conditions

$$P\frac{\partial F}{\partial X_1} - w_1 = 0$$

$$\cdots$$

$$P\frac{\partial F}{\partial X_N} - w_N = 0.$$

This is similar to our results from the consumer problem, but now we lack the multiplier. In the consumer problem, the (endogenous) multiplier converts dollars into utility, but since output is sold at price P, that is the value of a unit of output; the price replaces the multiplier. Rearranging, we can write the marginal product in terms of prices:

$$\frac{\partial F}{\partial X_1} = \frac{w_1}{P}$$

$$\cdots$$

$$\frac{\partial F}{\partial X_N} = \frac{w_N}{P}.$$

Further, we can solve this system of equations to get input-demand functions

$$X_1 = X_1(w_1, \ldots, w_N, P)$$
$$\cdots$$
$$X_N = X_N(w_1, \ldots, w_N, P).$$

We can also derive a resulting supply function:

$$Y = Y(w_1, \ldots, w_N, P).$$

Recall that before we wrote $P = MC$. The system of first-order conditions derived in this section is identical to setting price equal to marginal cost. Just rearrange the first-order condition to get that $\dfrac{w_i}{\frac{\partial F}{\partial x_i}} = P$. This is

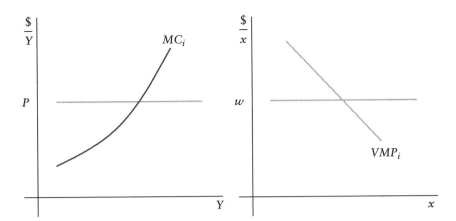

Figure 10-4: Setting wage equal to the value of marginal product is the same as setting price equal to marginal cost.

saying that dollar cost per marginal unit of output for increasing that input equals the output price. See also Figure 10-4.

Asking "How much input should I use?" is identical to asking "How much output should I produce?"—we're simply using a different perspective.

COST MINIMIZATION

The new problem has two steps. In step 1, minimize cost for a given level of output. In step 2, pick the level of output to maximize profits.

Step 1

$$\min_{X_1, \ldots, X_N} \sum_{i=1}^{N} X_i w_i \quad s.t. \; F(X_1, \ldots, X_N) = Y.$$

This problem, in turn, yields the cost function $C(w_1, \ldots, w_N, Y)$ This is like solving for Hicksian demand functions. Now the constraint, instead of being a utility function, is the production function. Importantly, this time we can tell empirically if a firm is holding output constant. As with the expenditure function, taking the derivative of the cost function with respect to the wage yields demand:

$$\frac{\partial C}{\partial w_i} = X_i(w_1, \ldots, w_N, Y).$$

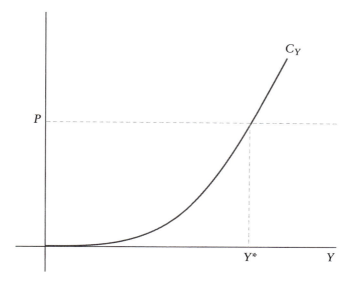

Figure 10-5: The firm chooses the level of output so that price equals marginal cost.

These are called *conditional factor demands* because they are conditional on output. This is the end of step 1.

Step 2

$$\max_Y PY - C(w_1, \ldots, w_N, Y).$$

Sometimes the problem will start here. Just remember that this function C implicitly assumes a production function. The first-order condition is

$$P - \frac{\partial C}{\partial Y} = 0,$$

which is again the fact that $P = MC$. This is shown in Figure 10-5.

Recall in the consumer problem we could go back and forth between the Marshallian and Hicksian demand curves using the Slutsky equation. Unconditional factor demands are different from Marshallian demand because firms can purchase more input by being efficient producers, but individuals cannot purchase more goods by being efficient producers of utility.

As an aside, think about why this problem behaves well. For instance, why is marginal cost increasing? Equivalently, why does $F(X_1, \ldots, X_N)$

exhibit decreasing returns to scale?[3] Why can't a firm just replicate current inputs to double output? It might be difficult to replicate exactly what exists. No set of workers for hire will be exactly the same as currently hired workers. Even if such workers could be found, it might be hard to find them. Similarly, even if the same inputs can be found, some amount of time is required for replication. Once double the operation is occurring, firms also tend to require additional administration, raising marginal cost.

A common practice in applied industrial organization is to look at the expenses of a business and declare that materials and production employees are marginal costs but that management, space, and other "fixed" expenses are not marginal costs. Naturally, these studies find that the business charges more than the narrowly defined marginal costs, but this may not tell us that price exceeds marginal cost. Rather, the business recognizes that engaging in more production will hasten the date, or increase the probability, that management, space, and other so-called fixed expenses have to be expanded. These are marginal costs too.

THE FIRM'S SLUTSKY EQUATION

Though we do not have the Slutsky equation, there is an equation relating unconditional and conditional factor demands. Recall,

$$\frac{\partial C(w_1, \ldots, w_N, Y)}{\partial w_i} = X_i(w_1, \ldots, w_N, Y), \quad \text{and} \quad P = \frac{\partial C(w_1, \ldots, w_N, Y)}{\partial Y}.$$

Totally differentiate with respect to w_i, allowing Y to adjust (that's what we mean by unconditional factor demand), to get

$$\frac{\partial X_i^{UC}}{\partial w_i} = \frac{\partial^2 C}{\partial w_i^2} + \frac{\partial^2 C}{\partial w_i \, \partial Y} \frac{dY}{dw_i}$$

$$0 = \frac{\partial^2 C}{\partial w_i \, \partial Y} + \frac{\partial^2 C}{\partial Y^2} \frac{dY}{dw_i}$$

where $\dfrac{\delta X_i^{UC}}{\delta W_i}$ is a partial derivative because it holds constant all other input prices but is unconditional because it allows the level of output to respond to w_i. Eliminating $\dfrac{dY}{dw_i}$ yields

$$\frac{\partial X_i^{UC}}{\partial w_i} = \frac{\partial^2 C}{\partial w_i^2} - \left(\frac{\partial^2 C}{\partial w_i \, \partial Y}\right)^2 \bigg/ \frac{\partial^2 C}{\partial Y^2}.$$

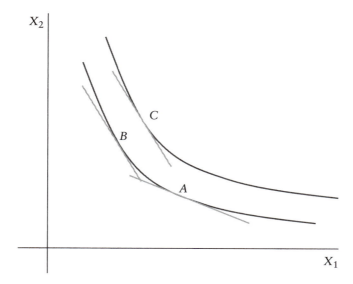

Figure 10-6: Consider an increase in the price of X_1. The substitution effect leads the firm to shift from A to B. Though total costs have gone up, in this case marginal cost went down so output is increased—the scale effect leads the firm to shift from B to C.

This is the firm's version of the Slutsky equation. We know that $\dfrac{\partial^2 C}{\partial w_i^2} \leq 0$ because the cost function is concave in prices. We also know that $\dfrac{\partial^2 C}{\partial Y^2} > 0$, since this is the change in marginal cost as we increase output. If marginal cost is falling, we would be finding a minimum, so it must be rising. Now the squared term cannot be negative, so we can conclude that X_i is not increasing in w_i. Unlike in the consumer problem, where we have Giffen goods, input demand cannot increase in its factor price.

Now what can we say about output? The sign is actually ambiguous. The firm may increase output as w_i increases if it can better avoid higher cost by increasing output. If two inputs are substitutable, for example, and the less productive input becomes more expensive, the firm may increase total output when it shifts to using the more productive input.

In the case shown in Figure 10-6, the total (and average) cost for a given level of output has still gone up, but the marginal cost of output at that level may have decreased, leading to an increase in output. However, the case of a factor price reducing marginal cost is rare. The factors associated with these cases are called *inferior factors*.

TWO-INPUT PRODUCTION

Often, we write output as a function of capital and labor:

$$Y = F(K, L)$$

K has a capital or purchase price and a rental price, R. The capital price measures the value over the life of the unit of capital. The rental price, on the other hand, measures the price of current use. For example, renting computers that would cost \$1 million to buy is much more expensive than renting factories that would cost \$1 million to buy, because the factories will produce their \$1 million in value over decades, whereas computers become obsolete quickly and must produce their value quickly.

Labor has the rental price w. Paying the wage gives the firm access to a person's labor for a specified amount of time. So firms will solve

$$\max_{K,L} PF(K, L) - wL - RK.$$

In this case, we get the same first-order conditions as before:

$$P\frac{\partial F}{\partial L} = w$$

$$P\frac{\partial F}{\partial K} = R,$$

as illustrated in Figures 10-7a and 10-7b. One useful result from these conditions is that we might measure marginal cost as the ratio of either factor's rental price to its marginal product—either ratio should give the same answer.

When thinking about capital, it is important to think about the short run versus the long run. A common view in production is that capital is fixed in the short run but variable in the long run. Labor has traditionally been viewed as much more flexible. These notions have become less useful over time. A law firm trying to expand, for example, will have a much harder time building an integrated labor force than finding office space, computers, and so on.

Now, consider a reduction in the wage rate, from w to $\hat{w}$. This shift is depicted in Figure 10-8. In the short run, labor will increase. In the long run, we will need to think about the cross partial derivatives of the

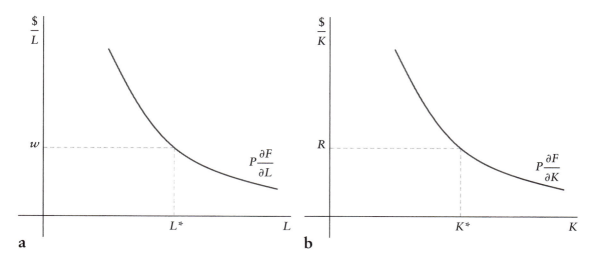

Figures 10-7a and 10-7b: Firms optimally set the value of marginal products equal to the input prices. Note that F depends on both K and L, so its derivative might as well. That is, the value marginal product curve in the left graph may change as K changes, and similarly in the right graph.

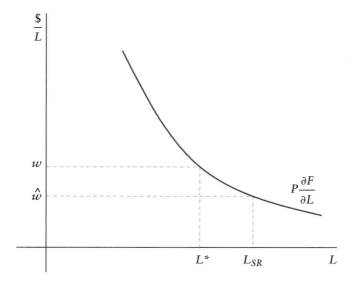

Figure 10-8: A decrease in the wage increases labor in the short run.

production function—whether labor and capital are complements or substitutes. Suppose that $F_{LK} > 0$; that is, that the marginal productivity of labor is increasing in the amount of capital. Then since we have increased labor in the short run, we will optimally increase capital over time. This will have additional feedback effects with labor, since increasing capital will further make labor more productive. Eventually, we approach a long-run equilibrium with a higher level of labor and capital, as illustrated in Figures 10-9a and 10-9b.

Now, what about the case where $F_{LK} < 0$? This means that capital and labor are substitutes. In this case, the firm would choose optimally to *reduce* capital, since it can substitute the cheaper input of labor. The feedback effects still exist, but they serve to drive capital lower and increase labor more. Again, labor is more elastic in the long run than in the short run. Since capital is a complement to labor if and only if labor is a complement to capital, the long-run effect always magnifies the short-run effect.

The same is often (though not always) true for consumption. If the price of gas goes up, in the short run people will drive less. If cars and gasoline are complements, in the long run they will buy fewer cars and drive even less. If biking and driving are substitutes, then in the short run people will bike more and in the long run more people will buy bikes and biking will increase even more. Either way, the long-run effect is larger than the short-run effect.

This idea of feedback effects, however, deserves some mathematical discussion. These effects converge because of the second-order condition of the problem. We have assumed decreasing marginal productivity; so, for a generic objective function $f(X, Z)$, we know that $f_{XX} < 0$ and $f_{ZZ} < 0$.[4] For profit maximization, we also need that $f_{XX}f_{ZZ} - f_{XZ}^2 > 0$. The simple approach to solving this problem we have discussed by mentioning feedback effects is akin to solving the maximization problem in two steps. Step 1 is to maximize over one variable (say, Z) for a fixed value of the other

$$G(X) = \max_Z f(X, Z).$$

We want to show that G is concave. At the optimal Z^*, we have $f_Z = 0$, $f_{ZZ} < 0$. Total differentiation of the equality $f_Z = 0$ gives $f_{ZX} + f_{ZZ}\dfrac{dZ}{dX} = 0$, which gives us $\dfrac{dZ}{dX} = -f_{ZX}/f_{ZZ}$. We also have

$$G_X = \frac{df(X, Z^*(X))}{dX} = f_X + f_Z \frac{dZ^*}{dX} = f_X.$$

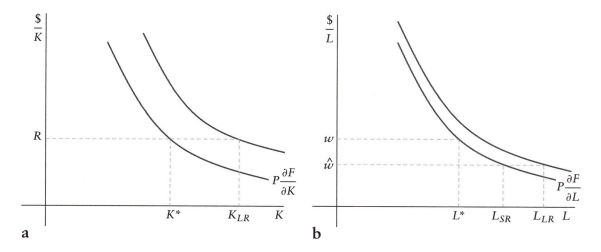

Figures 10-9a and 10-9b: Since we have assumed labor and capital are complements, a decrease in the wage increases capital in the long run. As the firm increases capital, it will also seek to increase labor further. Thus, the economy eventually approaches the depicted long-run equilibrium.

Total differentiation gives $G_{XX} = f_{XX} + f_{XZ}\dfrac{dZ}{dX}$. We can plug in for $\dfrac{dZ}{dX}$ to achieve that

$$G_{XX} = f_{XX} - \frac{f_{XZ}^2}{f_{ZZ}} < 0.$$

This determinant tells us that once we optimize over one variable as a function of the second, the problem is still concave in the second variable. In step 2, maximize over the other variable, and repeat this process.

The "Slutsky" equation for the firm discussed above can also be interpreted as a sequential optimization problem: get the right factor amounts for a given output, and then get the right output. Recall that we can access the labor demand function from the cost function, in the same way that we could derive Hicksian demand functions from the expenditure function. That is, $L = \partial C(w, R, Y)/\partial w$. Totally differentiating this, allowing output to adjust with w, yields that $\dfrac{\partial L^{UC}}{\partial w} = \dfrac{\partial^2 C}{\partial w^2} + \dfrac{\partial^2 C}{\partial w \partial y}\dfrac{dY}{dw}$. Similarly, use the first-order condition from step 2 of the cost version of the firm problem: $P = \dfrac{\partial C(w, R, Y)}{\partial Y}$. Again, totally differentiating yields $0 = \dfrac{\partial^2 C}{\partial w \, \partial y} + \dfrac{\partial^2 C}{\partial y^2} \dfrac{dY}{dw}$. Combining these

two results to eliminate the term $\dfrac{dY}{dw}$ gives

$$\frac{\partial L^{UC}}{\partial w} = \frac{\partial^2 C}{\partial w^2} - \left(\frac{\partial^2 C}{\partial w \, \partial y} \right)^2 \bigg/ \frac{\partial^2 C}{\partial y^2}.$$

SUBSTITUTION AND SCALE EFFECTS ON FACTOR DEMAND

We have repeatedly expressed the Slutsky equation for the firm in terms of own-price. That is, we have looked at the effect on the demand for input X_i if we increase the price of X_i. In the same exercise we have done twice now, one can show more generally that

$$\frac{\partial X_i^{UC}}{\partial w_j} = \frac{\partial^2 C}{\partial w_i \, \partial w_j} - \left(\frac{\partial^2 C}{\partial w_i \, \partial Y} \frac{\partial^2 C}{\partial Y \, \partial w_j} \right) \bigg/ \frac{\partial^2 C}{\partial Y^2}.$$

The result obtained before is the special case $j = i$. Now, consider the two

terms on the right-hand side of the equation. The left term, $\dfrac{\partial^2 C}{\partial w_i \, \partial w_j}$, is

called the *substitution effect*. The substitution effect tells us how much input j's price induces a shift along the isoquant toward input i. Since the cost function is the result of firm optimization, it carries with it the assumptions of the underlying production function used in step 1 of the problem.

The right term, $\left(\dfrac{\partial^2 C}{\partial w_i \, \partial Y} \dfrac{\partial^2 C}{\partial Y \, \partial w_j} \right) \bigg/ \dfrac{\partial^2 C}{\partial Y^2}$, is called the *scale effect*. It

is the effect of the factor price on factor demand through changes in

the scale of production. Because $\dfrac{\partial C}{\partial w_i}$ is the demand function X_i,

$\dfrac{\partial^2 C}{\partial w_i \, \partial Y} = \dfrac{\partial X_i}{\partial Y}$. X_i is an *inferior input* if and only if $\dfrac{\partial^2 C}{\partial w_i \, \partial Y} < 0$. Note

that a factor price cannot reduce cost, but it can (and does in the inferior-input case) reduce marginal cost. Lower marginal cost means that output can expand when one of the factors gets more expensive. Maybe a firm is using shovels to do small digging projects. But then shovels get more expensive, so the firm switches to digging with an excavator machine. As long as the firm has the machine, the firm digs more.

Inferior inputs are not that common, and we now see from the firm's Slutsky equation that additional production- or cost-function restrictions are needed to guarantee that no inputs are inferior. Moreover, with $i=j$ the Slutsky equation's scale effect term must be negative even though the direction of the scale effect is ambiguous—inputs can be normal or inferior. With a normal input, scale and factor demand move together and the factor price reduces scale. With an inferior input, scale and factor demand move in opposite directions but the factor price increases scale. Either way, the factor price reduces factor demand through the scale effect. The scale effect of a factor's own price always reinforces the substitution effect.

ACQUIRED COMPARATIVE ADVANTAGE

We began this chapter by assuming that production factors just happened to be different in ways that created comparative advantage. But the marketplace gives people an incentive to become different, to strengthen their comparative advantage. This result was explored in part of the introduction of this book, and we return to it now that we have invested in the analytics of comparative advantage. We go back to the simplest example of comparative advantage and then add human capital acquisition to it.

Think about a simple world with two tasks, A and B. An individual has human capital for those tasks, H_A and H_B. Whatever task is chosen, the individual is paid a wage per unit human capital: w_A or w_B, as appropriate. This will mean total income for an individual from task A is $Y_A = w_A H_A$ and from task B is $Y_B = w_B H_B$. If an individual can only choose one task, the individual will choose to earn income

$$Y = \max(w_A H_A, w_B H_B)$$

So, the individual picks task A if $w_A H_A > w_B H_B \Leftrightarrow \dfrac{w_A}{w_B} > \dfrac{H_B}{H_A}$. This is comparative advantage because the choice depends on the relative amounts of human capital that the individual has, not the absolute amount.

We illustrate the choice in the $[H_A, H_B]$ plane by drawing a task-indifference ray showing all of the configurations of human capital that

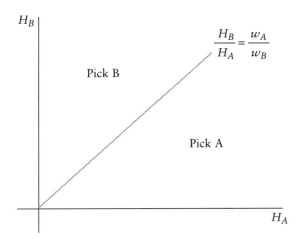

Figure 10-10: Supply and demand will rotate the task-indifference ray until the right number of workers is in each task.

someone could have and be indifferent between the two tasks. See Figure 10-10.

There is demand for tasks A and B, which in equilibrium has to match up with the available human capital and the aforementioned incentives for workers to choose one task rather than the other. This happens with wage adjustments. If there were a lot of demand for A, then Figure 10-10's task-indifference ray has to be steep so that lots of workers choose task A and few choose B. In other words, w_A/w_B would be greater than 1.

Now, assume we have reached the equilibrium, so that w_A/w_B reflects market supply and demand. Then for any point on the line, every person directly below and directly left must be earning the same income. See the dashed lines in Figure 10-11. This is because each person on the dashed line above the task-indifference ray has the same level of H_B and his or her H_A does not matter because it is not used. Each person on the dashed line below the task-indifference ray has the same level of H_A and his or her H_B does not matter because it is not used. Let's call the union of the two dashed lines an indifference curve for the worker.

Now, let's allow all agents to choose their human capital. For example, an agent is considering whether to be a good plumber versus a good carpenter. The opportunity set for human capital could have an interesting shape, as depicted in Figure 10-12. Consider the point associated with the maximum level H_B. As it is depicted, this person will have some

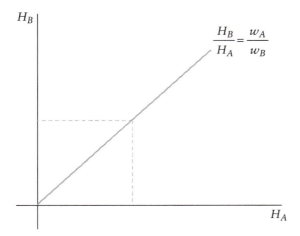

Figure 10-11: A worker indifference curve: each person on the dotted line earns the same income.

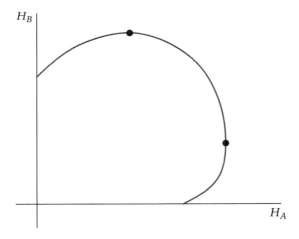

Figure 10-12: The opportunity set for selecting human capital. The agent with maximum human capital for task A still has positive human capital for task B.

positive level of H_A. This reflects an underlying story that some of tasks A and B require some of the same abilities. Thus, if the agent chooses to be a good plumber, that doesn't mean that the agent ends up with zero human capital as a carpenter.

Note further that in this graph, the economically relevant region of the opportunity set lies between the two points, and we can erase the parts of the curve close to the axes because no one would choose a human

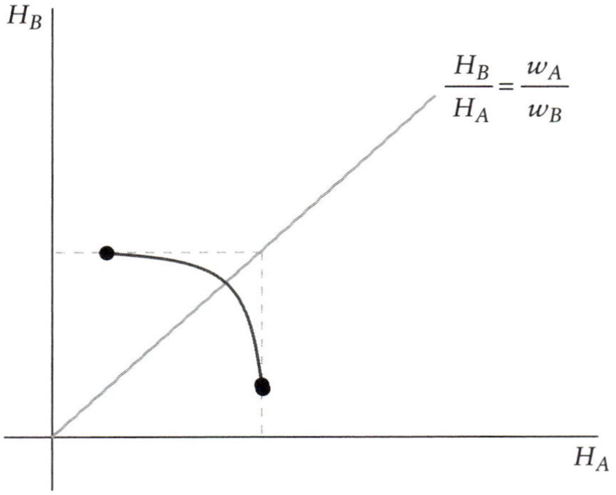

Figure 10-13: Specialization. Agents maximize their human capital at task *A* or task *B*.

capital pairing left of the top point or below the right point. On the erased regions, the agent could be better at both tasks!

Now let's put the opportunity set together with the worker's indifference curves, as in Figure 10-13. We can even have everyone identical in the sense that they all have the same opportunity curve to choose from. Nevertheless, specialization is optimal behavior. Being equally good at tasks *A* and *B* is worse than being very good at just one task because you have acquired a lot of human capital that you do not use.

We started this picture by indicating the types of workers (i.e., configurations of human capital) who are indifferent between the two tasks. But now we have shown that people will not choose to be those types of workers. Because human capital is acquired, it is not an equilibrium for people to be indifferent between the two tasks.[5]

The equilibrium requires that both tasks be performed, so some people specialize in *A* and others specialize in *B*. People who are identical in the sense of having the same opportunities open to them end up being different.

One might say that it is a coin flip exactly who goes toward task *A* and who goes toward task *B*, and we would agree if people were precisely identical. But in reality, people have somewhat different opportunities open to them: in Figure 10-13, that means somewhat different opportunity curves. Some of the opportunity curves may be relatively steep and others relatively flat. Then just a small difference among people in the slope of the curve will decide who specializes in what. Specialization in the marketplace turns small differences into large differences.

Gary Becker revolutionized labor economics by showing how so many of the differences among workers are acquired (Becker 1964; Becker and Murphy 1992). The differences did not just happen independent of supply and demand considerations.

Chapter 11

The Industry Model

PROPERTIES OF THE INDUSTRY MODEL

We begin with the industry model of demand and assume constant returns to scale at the industry level.[1] Constant returns to scale is a somewhat problematic assumption at the firm level, because if the price is above cost, it is optimal to produce an infinite amount of output; if the price is below cost, it is optimal not to produce; and if the price is equal to cost, output is indeterminate. Because firm-level equilibria are often indeterminate, it is often necessary to move to the industry level to begin considering equilibria. When we assume the industry exhibits constant returns to scale, we are *not* necessarily assuming that firms have constant returns to scale. Each firm in the industry could have diminishing returns to scale, but expanding the industry as a whole can exhibit constant returns to scale—replication may be feasible on the industry level. So, assuming constant returns at the industry level encapsulates two cases: (1) firms in the industry exhibit constant returns to scale, or (2) firms in the industry exhibit diminishing returns, which in the aggregate are well approximated by a constant-returns-to-scale model of the industry.

We further assume the two-input case. Then constant returns to scale means

$$F(tL, tK) = tF(L, K).$$

For our conditional factor demand functions, constant returns to scale will imply

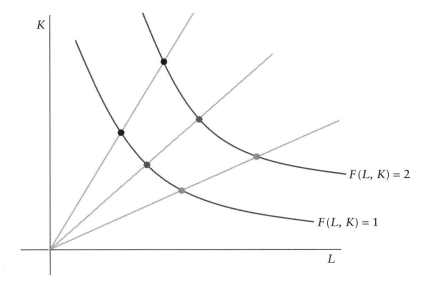

Figure 11-1: Doubling inputs is associated with doubling output. Further, at each of the same colored points, the slopes of the isoquants are the same.

$$L^*(w, r, Y) = YL^*(w, r, 1)$$
$$K^*(w, r, Y) = YK^*(w, r, 1).$$

The derivatives are therefore homogeneous of degree zero (see also Figure 11-1):

$$\frac{\partial F(tL, tK)}{\partial L} = \frac{\partial F(L, K)}{\partial L}$$
$$\frac{\partial F(tL, tK)}{\partial K} = \frac{\partial F(L, K)}{\partial K}.$$

We can also make statements about the cost function:

$$C(w, r, Y) = YC(w, r, 1).$$

For the cost function, constant returns to scale implies that the cost of producing Y units is the same as the cost of producing one unit Y times, taking the derivative, $\dfrac{\partial C(w, r, Y)}{\partial Y} = C(w, r, 1)$, so that marginal cost is constant. In sum, constant returns to scale says that relative factor prices determine the ratio of K to L, and the amount of output will determine

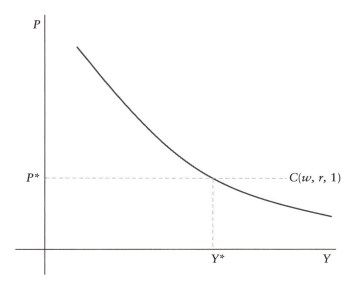

Figure 11-2: Optimal output is determined by the intersection of the constant marginal cost curve, which equals the price, and consumer demand.

the levels of K and L required. Thus, CRS gives a convenient decomposition of relative input use and the level of output.

It remains to determine how optimal output is pinned down. Recall the condition $P = MC$. In this context, that means $P = \dfrac{\partial C(w, r, Y)}{\partial Y}$ $= C(w, r, 1)$. We also have market clearing, $Y = D(P)$, so that the amount of output equals the amount of the good that consumers demand, as illustrated in Figure 11-2.

Supply is perfectly elastic, so the supply side determines the price. The quantity is then determined by how many consumers are willing to purchase the good at the determined price.

As an aside, note that for most industries, this is a good way to think about the world. Prices are determined on the supply side by what it costs to produce the product. The quantity we consume is determined on the demand side by how many goods we want to consume at the given price.

THE SUPPLY-DEMAND PERSPECTIVE ON INDUSTRY BEHAVIOR

The supply and demand perspective has a lot to say about important issues ranging from inequality (Katz and Murphy 1992) to the war on

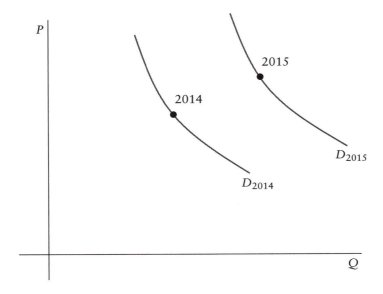

Figure 11-3: Since the demand curve intersects the equilibrium point, the demand curve must have shifted outward from 2014 to 2015. Moreover, the change in quantity is greater than the change in demand for each equilibrium price.

drugs (chapter 12 of this book) to the business cycle (Mulligan 2012). Although it is something that should be mastered in undergraduate studies, even the experts sometimes get confused about what the supply and demand perspective has to say.[2]

Consider the case of a change in price and quantity between the years 2014 and 2015. For the moment we're not going to necessarily assume that supply is perfectly elastic. In 2015, both the equilibrium price and quantity are higher than they were in 2014. Because demand curves must pass through both points, we know that demand must be higher in 2015 than it was in 2014. We also know that the change in demand (ΔD, measured in the quantity dimension as a percentage of the initial quantity) is greater than the change in quantity—this is evident from Figure 11-3.[3] The same graph for supply would imply that the change in supply (ΔS, also measured in the quantity dimension as a percentage of the initial quantity) is less than the change in quantity. In summary, if we let ΔP and ΔQ denote the percentage equilibrium changes in price and quantity, respectively, then $\Delta D > \Delta Q > \Delta S$ when $\Delta P > 0$ and the opposite when $\Delta P < 0$.

We can think about this example in terms of elasticities as well. Note that the quantity we observe equals the demand evaluated at market prices, $Q = F(P, Z)$, where Z refers to anything that might shift the demand curve, such as the price of another good. Thus, changes in quantities can be written in terms of changes in P and changes in Z according to $d \ln Q = \dfrac{1}{Q}(F_P dP + F_Z dZ) = F_P \dfrac{P}{Q}\dfrac{dP}{P} + \dfrac{F_Z dZ}{Q}$. We will write this as $\Delta Q = \Delta D + \epsilon^D \Delta P$, where $\Delta D = F_Z dZ / Q$, $\Delta Q = d \ln Q$, and $\Delta P = d \ln P$.

We have decomposed the change in quantity into a shift in the demand curve and a movement along the demand curve.

The condition $\Delta Q = \Delta D + \epsilon^D \Delta P$ implies that $\Delta D = \Delta Q - \epsilon^D \Delta P$. Similarly, on the supply side, $\Delta Q = \Delta S + \epsilon^S \Delta P$, which implies that $\Delta S = \Delta Q - \epsilon^S \Delta P$. In both cases, ϵ denotes an elasticity of quantity with respect to price. In this example, we know ΔQ and ΔP; so, given also the elasticities of supply and demand, we could calculate exactly what the shift in demand and supply had to be.

Note that the equilibrium price and quantity are on both the supply curve and the demand curve. An equilibrium quantity must simultaneously be the quantity that sellers want to sell and the quantity that buyers want to buy. That is why we are keeping track of both a ΔD equation and a ΔS equation. This is a simple point, but people often forget one side of the market or the other.

Because ΔQ is shared in both equations, we can eliminate it and solve for ΔP as a function of the demand and supply shifts:

$$\Delta D + \epsilon^D \Delta P = \Delta S + \epsilon^S \Delta P$$
$$\Rightarrow \Delta P = \frac{\Delta D - \Delta S}{\epsilon^S - \epsilon^D}.$$

This says that the percentage change in price is equivalent to the change in excess demand divided by the "sum" of the elasticities, because $\epsilon^D < 0$. Similarly,

$$\Delta Q = \frac{\epsilon^S \Delta D - \epsilon^D \Delta S}{\epsilon^S - \epsilon^D}.$$

Thus the percentage change in quantity is a weighted average of the percentage change in demand and the percentage change in supply.

FOUR INGREDIENTS OF THE INDUSTRY MODEL

In our constant returns model of the industry, we can incorporate our additional equilibrium conditions $L = \dfrac{\partial C(w, r, Y)}{\partial w}, \quad K = \dfrac{\partial C(w, r, Y)}{\partial r},$ and $Y = F(L, K)$. To summarize, the industry model therefore has four ingredients:

1. $P = MC$
2. $Y = D(P)$

3. $L = \dfrac{\partial C(w, r, Y)}{\partial w}$ and $K = \dfrac{\partial C(w, r, Y)}{\partial r}$

4. $Y = F(L, K)$

We can consider each of these in turn. The first equation corresponds, in the constant returns case, to $P = C(w, r, 1)$. Totally differentiating, then $dP = C_w(w, r, 1)dw + C_r(w, r, 1)dr$. Because of constant returns to scale, then $dP = \dfrac{L}{Y}dw + \dfrac{K}{Y}dr$. Multiply and divide appropriately to convert changes to percentages: $\dfrac{dP}{P} = \dfrac{wL}{PY}\dfrac{dw}{w} + \dfrac{rK}{PY}\dfrac{dr}{r}$. This implies that $\Delta P = S_L \Delta w + S_K \Delta r$, where S denotes factor spending as a share of revenue and Δ denotes a percent change.[4] So the change in the price of output is a share-weighted average of the changes in input prices. As long as price equals marginal cost, this equation must be satisfied.

The second equation implies that $\Delta Y = \epsilon^D \Delta P$.

The third equation makes a new concept relevant: the elasticity of substitution, typically denoted σ. This measures how much factor inputs respond to changes in factor prices along the isoquant. Because of our previous discussion about the effects of a CRS assumption on isoquants (see Figure 11-1), we know that moving along a ray starting at the origin to different isoquants will maintain the same elasticity of substitution. More precisely, we will define $\sigma > 0$ to mean

$$\Delta \frac{L}{K} = \sigma \Delta \frac{r}{w}$$

There is a negative relationship between the ratio of factor quantities and the ratio of factor prices (recall that r is K's price, and w is L's). This condition implies that

$$\Delta L - \Delta K = \sigma(\Delta r - \Delta w).$$

The fourth equation can be treated in the same way we treated the first equation. $Y = F(L, K)$ implies that $dY = \dfrac{\partial F}{\partial L} dL + \dfrac{\partial F}{\partial K} dK$. This means that

$$\frac{dY}{Y} = \frac{P\dfrac{\partial F}{\partial L}}{PY} \frac{dL}{L} + \frac{P\dfrac{\partial F}{\partial K}}{PY} \frac{dK}{K}, \text{ which yields } \Delta Y = S_L \Delta L + S_K \Delta K.$$

INDUSTRY ELASTICITY OF LABOR DEMAND

The first confounding element one might consider when contemplating how a change in the wage affects labor demand is capital. For now, assume $\Delta r = 0$. This is typically a long-run assumption. Then $\Delta P = S_L \Delta w$, $\Delta Y = \epsilon^D \Delta P$, $\Delta L - \Delta K = -\sigma \Delta w$, and $\Delta Y = S_L \Delta L + S_K \Delta K$. When the wage changes, we will see changes in price, output, labor, and capital, so we have four equations to solve for four unknowns. Using the first two, we can simplify to achieve

$$\Delta Y = S_L \epsilon^D \Delta w.$$

This is the scale effect. An increase in the wage here drives output down. Now rewrite the fourth equation as $\Delta Y = \Delta L + S_K(\Delta K - \Delta L)$. This implies that $\Delta L = \Delta Y + S_K(\Delta L - \Delta K)$. This means that the change in labor is going to be the scale effect *and* the substitution effect. We can use the third equation to plug in for $\Delta L - \Delta K$ to achieve $\Delta L = S_L \epsilon^D \Delta w + S_K (-\sigma \Delta w)$. The first term is the scale effect, and the second is the substitution effect. This can be rewritten to yield

$$\Delta L = (S_L \epsilon^D - S_K \sigma)\Delta w.$$

The more elastic is output demand, the more labor will decline. The more substitutable labor and capital are, the more labor will fall. These two are part of *Marshall's Law* (Marshall 1890). A third aspect of Marshall's law stated that increasing labor's share would drive labor further downward; the issue with this argument is that S_L and S_K are, of course, related; so increasing S_L also reduces the ability to substitute capital for labor (Hicks later corrected this law,[5] noting that it additionally requires ϵ^D to have greater magnitude than σ).

Now consider, as an exercise, the short-run elasticity of labor demand. That is, hold capital fixed so that $\Delta K = 0$. This will give the equations $\Delta P = S_L \Delta w + S_K \Delta r$, $\Delta Y = \epsilon^D \Delta P$, $\Delta L = \sigma(\Delta r - \Delta w)$, and $\Delta Y = S_L \Delta L$. Solving these equations in the same way as before will yield the short-run elasticity of demand for labor.

ARE LABOR AND CAPITAL COMPLEMENTS OR SUBSTITUTES?

This can be restated: in the long run ($\Delta r = 0$), does $\Delta w > 0$ imply $\Delta K > 0$ or $\Delta K < 0$? In the short run ($\Delta K = 0$), does $\Delta w > 0$ imply $\Delta r > 0$ or $\Delta r < 0$?

Recall that the scale effect works in the following way: $\Delta w > 0 \Rightarrow \Delta P > 0 \Rightarrow \Delta Y < 0$, so the scale effect is always pushing in the direction of less labor and capital. The substitution effect means $\Delta w > 0 \Rightarrow \Delta \dfrac{K}{L} > 0 \Rightarrow \Delta K > 0$, so K is driven upward. So the question boils down to: is the elasticity of demand or the elasticity of substitution more important?

In the long run, using $\Delta L - \Delta K = -\sigma \Delta w$ and the formula for the change in labor, we have

$$\Delta K = (S_L \epsilon^D - S_K \sigma)\Delta w + \sigma \Delta w = S_L(\epsilon^D + \sigma)\Delta w.$$

In the short run, $\Delta K = 0$, so the equilibrium equations are $\Delta P = S_L \Delta w + S_K \Delta r$, $\Delta Y = \epsilon^D \Delta P$, $\Delta L = \sigma(\Delta r - \Delta w)$, and $\Delta Y = S_L \Delta L$. These give that $S_L \sigma(\Delta r - \Delta w) = \epsilon^D(S_L \Delta w + S_K \Delta r)$, so $(S_L \sigma - S_K \epsilon^D)\Delta r = (\epsilon^D + \sigma)S_L \Delta w$. Rearranging gives

$$\Delta r = \Delta w \frac{(\epsilon^D + \sigma)S_L}{(S_L \sigma - S_K \epsilon^D)}.$$

Since $\epsilon^D < 0$, the denominator is positive. When $(\epsilon^D + \sigma) > 0$, the substitution effect dominates the scale effect, so that the price of capital rises in the short run and amount of capital rises in the long run.

Consider a subsidy to capital in the automobile industry. Will that lead to more or less labor in the industry? Assume a closed economy, where the country produces and consumes all of its cars. The scale effect is likely to be smaller in this closed economy case because there are fewer substitutes in the output market. Now suppose the state of Illinois decides to subsidize capital in the automobile industry. This would tend to make the scale effect bigger in Illinois, since the demand for Illinois-produced cars is more elastic than the demand for U.S.-produced cars (because we can move across markets). The effect of a capital subsidy on employment is thus different depending on the level at which it is supplied, state versus national.

So the scale effect tends to make labor and capital complements, whereas the substitution effect tends to make them substitutes. Depending on the circumstance and horizon, the mix of substitution and scale effects can be different. This can make determining whether labor and capital are complements or substitutes a difficult question.

For another example, consider a pilot employed by a commercial airline. The pilot is a small share of the total cost for flights, so the demand for the pilot is fairly inelastic, in the sense that the scale effect will be small. But this means that the complementary shares are large, so that an increase in the price of pilots will drive airlines to use bigger planes and more fuel. Marshall (1890) neglected the fact that an input with only a small share increases the ability of the firm or industry to substitute other inputs, potentially increasing the magnitude of the substitution effect.

Chapter 12

The Consequences of Prohibition

THE REVENUE FROM DRUG SALES

To be concrete, let's consider a model of illegal drugs.[1] We want to think about there being a demand for drugs, and we want to initially assume they are legal and the industry is perfectly competitive and has constant marginal cost. Perfect competition is easy to relax, and the constant marginal cost seems to be a good assumption empirically. Now we will implement a prohibition on drugs and initially assume that there is no effect on demand. The key point will be that prohibition raises the cost of supplying drugs by forcing suppliers to do things they otherwise wouldn't. That is, they have to deviate from more efficient strategies. Raising the marginal cost raises the price. This reduces the quantity demanded. We'll assume demand is inelastic. For simplicity assume $\epsilon^D = -1/2$, and $\hat{p} = 4p_c$. Then $\hat{q} = \dfrac{1}{2}q_c$, since $q = Ap^{-\frac{1}{2}}$ is the constant-elasticity demand function. This means we are using twice as many resources as before to produce half as many drugs. Consumer expenditure is twice as much, going from E_c to $\hat{E} = 2E_c$ as in Figure 12-1. For this to have social benefits that exceed the costs, we must think that the externality from increasing consumption from $\hat{q}$ to q_c must be greater than the additional costs we incur *not* to produce that additional quantity.

Thus, this method of raising the real cost of consumption is inefficient unless it is believed that there is a huge negative externality from consumption of the additional drugs produced. But there are also other externalities imposed from making drugs illegal. For instance, drug dealers

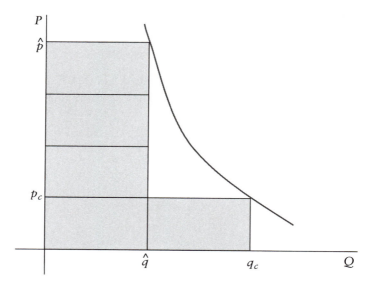

Figure 12-1: In this example, we use twice as many resources to produce half as many drugs.

are often very violent, and the effects of this violence affect people who are not drug dealers. Further, drug dealers may corrupt local officials around the world, or fail to pay income, sales, excise, and other taxes, which impose negative externalities that are broader in scope.

The legal status of drugs can affect demand by making purchases less convenient or more socially stigmatized. We can allow for this, while still using Figure 12-1, by reinterpreting the price shown in the figure as the full price. The full price is the sum of the money price paid to drug sellers and the additional inconvenience and stigma costs experienced by the consumer. We still have that prohibition delivers half the quantity at twice the cost, but now consumer expenditure on drugs measures only part of those costs. The willingness-to-pay-money schedule, not shown in the figure, is below the demand curve. Consumer expenditure and therefore industry revenue is measured as the equilibrium willingness to pay times the equilibrium quantity $\hat{q}$.

THE LEGALIZATION MULTIPLIER

Figure 12-1 shows a prohibition that fully eliminates legal activity E_c in the industry and replaces it with illegal activity that is twice as costly. Those extra resources come from elsewhere in the economy. Holding constant the supply of production factors to the total economy, and for simplicity assuming that the rest of the economy is legal, Figure 12-1 is showing that prohibition in this industry reduces legal activity outside of the industry just as much as it reduces it inside the industry, which itself is a lot. To the extent that legal activity is subject to income, sales, excise, and other taxes, the reallocation from legal to illegal is a negative externality of the prohibition itself in addition to the violence, corruption, etc. already mentioned.

The supply of labor to the total economy may not be constant because the prohibition affects productivity (producing few drugs with more resources), although the effect of productivity on labor may be small and of ambiguous sign because productivity has both income and substitution effects. As we show in later chapters, less productivity probably means less aggregate capital.[2]

It follows that the legalization of drugs would not only expand the legal drug sector but also other legal sectors. Legalization therefore has a lot in common with productivity growth in agriculture or some other industry with inelastic demand: the industry produces more while freeing up resources for the rest of the economy to also produce more.

HALF-HEARTED PROHIBITIONS ARE THE MOST COSTLY

Note that, while this analysis has been largely against the idea of a prohibition, prohibitions can be beneficial in some situations. If demand becomes elastic at low quantities (as it must for any demand curve with a choke point)[3] and enforcement is effective enough that quantity is almost fully reduced to zero, then even though the per-unit costs of the goods still produced will be exceptionally high, the total costs will still be low because they are incurred for so few goods. With this type of demand curve, prohibition is only a bad policy when it is not particularly effective. When it's not effective, prohibition imposes significant costs on suppliers and society at large. It is only worth it in these cases if there is a strongly negative effect of the additional goods being available.

In fact, we may simply want to exterminate an industry where we would not want merely a modest reduction. See Figure 12-2. Let's assume linear demand, which means that demand has a choke point near which demand is price elastic. Let's also assume a competitive industry-constant

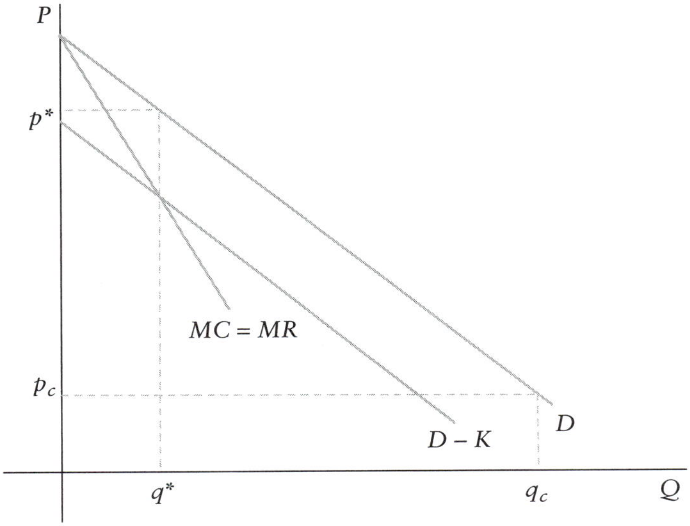

Figure 12-2: After accounting for the per unit externality *K*, it is optimal either to eradicate the industry or leave it at the competitive quantity and price. It is a minimum, in fact, at the pair (*q**, *P**) depicted.

marginal cost industry, which gives that total cost equals total revenue. Thus $P = D(Q) = C$ and $CQ = D(Q)Q$, where $D(Q)$ denotes the inverse demand curve. Government enforcement of a prohibition raises C and therefore P, and thereby indirectly determines consumption Q. Another way of thinking of this is that for any potential market size Q, the government can chose a level of enforcement of a prohibition such that the production cost (and therefore the price) is $D(Q)$. The socially optimal consumption balances the production costs CQ, and any external costs, with the benefits to consumers.

Because for any quantity, production costs are equal to industry revenue $D(Q)Q$, the marginal production cost of expanding quantity is equal to marginal industry revenue, $D(Q) + D'(Q)Q$. Recall from the usual monopoly model that an industry's marginal revenue curve is below the demand curve, as shown in Figure 12-2.

Now suppose K is the per unit externality. We shift the demand curve (i.e., private marginal benefit curve) down by K to get the marginal social benefit. Note that for all the outputs between 0 and q^*, where the marginal cost line intersects the marginal social value line $(D - K)$, marginal cost exceeds the marginal social value. In other words, at any quantity between 0 and q^*, society gains from further reductions in quantity.

Conversely, at any quantity between q^* and the (unregulated) quantity q_c, society gains from further increases in quantity. Thus, the quantity q^* where the two lines intersect minimizes social surplus.

So in this simple example, if we can set whatever price we want, we either want to do nothing and stay at the competitive output or completely eradicate the industry. Which one we choose to do depends on how large the externality is. We lose surplus on the first q^* units but then gain surplus as we move from q^* to q_c. We need to assess the net gain here. This process provides some intuition for why ineffective prohibitions can be very inefficient.

Chapter 13

A Price-Theoretic Perspective on the Core

So far we have looked at cases where equilibrium price and marginal cost coincide. We noted in the context of firm theory that this case is more applicable than it first appears because marginal costs should be interpreted broadly. Also, the equilibrium gap between marginal cost and price, if there is one, can sometimes be sufficiently constant that we get the correct comparative statics even when we take the gap to be zero. Because there are interesting behaviors for which a constant gap is not a satisfactory treatment, we use this chapter to show how a small adjustment to the previous analysis can add a lot of insight.

LOOKING FOR GAINS FROM TRADE: INDIFFERENCE CURVES FOR BUYERS AND SELLERS

Recall Figure 5-1, where we drew a consumer's indifference curve and demand curve on the same quantity-price diagram, based on the fact that an individual's demand curve indicates the optimal quantity purchased at each price. Each point on the demand curve, all the way up to the demand curve's choke point, has an indifference curve going through it.[1] The indifference curve going through the demand curve's choke point can be called the all-or-nothing demand curve. At any point on the all-or-nothing demand curve, the consumer is indifferent between bundles at prices along that curve and buying no units at the price where the demand curve intersects the y-axis. The consumer would rather not buy any units than purchase a bundle to the right of the

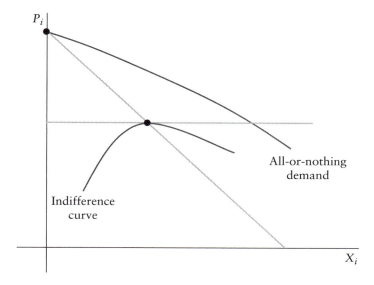

Figure 13-1: In (X_i, P_i) space, we can draw indifference curves. For any point along the all-or-nothing demand curve, for instance, the consumer is indifferent between purchasing a quantity at a price along that curve and purchasing no units at all at the price where his demand curve intersects the y-axis.

all-or-nothing demand curve. Figure 13-1 adds that to what is already shown in Figure 5-1.

Now think about a firm that faces a downward sloping demand curve. The firm is a monopolist, so it calculates the marginal revenue curve. For simplicity, we assume that marginal cost is constant. The familiar result is that the monopolist constrained by the demand curve sets quantity where marginal revenue equals marginal cost, as at Q^* in Figure 13-2.[2]

But we could get the same result by looking at the monopolist's "indifference curves"—more precisely, the isoprofit curves, which are points where profits are equal. The isoprofit curves corresponding to positive profits slope downward because getting the same profit at a lower price requires selling more units. Maximizing profits, taking the demand curve as a constraint, yields the profits associated with the seller's isoprofit curve that is tangent to the demand curve, which is also shown in Figure 13-2.

Now draw the consumer's indifference curve at the market equilibrium, (Q^*, P^*), and note that there is a region where both the consumer and the monopolist can be made better off. Point A in Figure 13-2 has been drawn in this region.

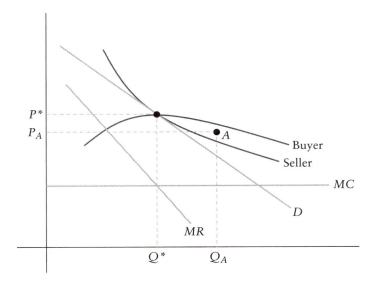

Figure 13-2: Both the monopolist and the consumer have an incentive to move from the monopoly solution to point A. Key features of point A are that the price is lower and quantity is higher than the market equilibrium, and the consumer is off the demand curve.

EXCLUSIVE DEALING, QUANTITY DISCOUNTS, AND OTHER MARKET OUTCOMES THAT ARE OFF THE MARSHALLIAN DEMAND CURVE

In the region between the isoprofit curve and the consumer's indifference curve, prices are lower than the monopolist's equilibrium price, quantity is higher than the equilibrium quantity, and the consumer is off his or her demand curve. That is, $P_A < P^*$, $Q_A > Q^*$, and $Q_A > Q^D (P_A)$. Note that both sides want to renege on this arrangement. After getting the lower price P_A, the consumer would rather purchase $Q^D (P_A)$. Similarly, after negotiating the higher quantity Q_A, the producer would rather receive revenues P^*Q_A. Thus, there needs to be some commitment here. One example of this scenario would be a negotiated discount, for instance.

This can be very good for consumers. Consumers are often willing to let grocery stores push them off their demand curves because this gives grocery stores more negotiating power with producers. In effect, they can make the consumer demand curve look more elastic by being able to push consumers off the demand curve in their stores.[3]

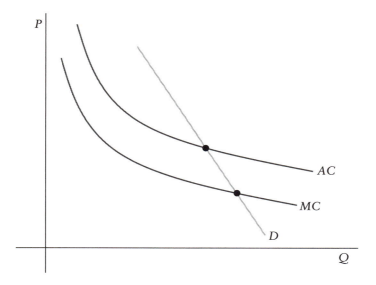

Figure 13-3: Marginal cost is below average cost, so the efficient outcome would mean firms are losing money on this good.

As explained by Klein and Murphy (2008), there is a similar reason for why so many fast-food chains serve Coke *or* Pepsi, rather than both. They have more negotiating power if they can tell Coke and Pepsi that they will lose all their business if they don't offer a deal. The ability to push people between Coke and Pepsi gives the fast-food chains this power. And typically, the price reduction achieved through this process compensates for the small losses consumers incur by being pushed around off their demand curves (i.e., getting more Coke than they'd like at the Coke-only restaurant and more Pepsi than they'd like at the Pepsi-only restaurants).

If buyers are heterogeneous, the monopolist constrained by demand may choose a price at which some of the buyers purchase nothing. Pepsi may choose a price that induces Coke lovers to drink no Pepsi. Here the all-or-nothing demand curve is relevant and shows how both the Coke lovers and Pepsi would be better off at a point below the all-or-nothing curve and above the Coke-lover demand curve. Such a point may be achieved when restaurants obtain a lower price for Pepsi by agreeing to serve Pepsi only. The Coke lover ends up drinking Pepsi when he dines at such restaurants, but pays less than he would at a restaurant that served both brands.

This perspective also helps us think about natural monopolies. Consider the structure and organization of the market as an outcome, not something that just happens. The problem with the natural monopoly model is that it does not consider the cost curves in a broader context. Consider the case in Figure 13-3, where marginal cost is below average cost.

The natural monopoly model says that we cannot achieve the efficient outcome because firms would be pricing below average cost. But what if we bundle this product with something else? Perhaps it would be possible to sell the good at marginal cost if it is possible to bundle it with another good where the firm makes enough to offset the cost. This is what the supermarket does. The supermarket doesn't just sell ketchup, which would have average cost above marginal cost as a stand-alone product. Instead, the supermarket bundles many items together, realizing some of the gains from trade that, according to the natural monopoly model, would be lost due to the cost structure of selling ketchup.

Chapter 14

Multiple-Factor Industry Model

REVIEW OF THE INDUSTRY MODEL

Recall the two-input model of industry demand. We ended up with four equations, which together implied

$$S_L \Delta L + S_K \Delta K = \Delta Y.$$

This is like a first-order approximation to the production function. In the Cobb-Douglas case, this *is* the production function. We also had

$$S_L \Delta w + S_K \Delta r = \Delta P.$$

This was a first-order approximation to the cost function, assuming constant returns to scale. The first equation did not assume constant returns to scale—if we did not have CRS, we would just have that $S_L + S_K \neq 1$. The cost-function approximation, on the other hand, will change in two significant ways. S_L and S_K will become marginal shares, and the equation will additionally become a function of Y, since changes in output will affect marginal cost when we do not have CRS. CRS really does two things: marginal factor shares become equal to average factor shares, because the industry uses inputs on the margin in the same way that it uses them on average, and the marginal cost becomes independent of output.

Note that homothetic production retains the feature that the isoquants increase radially out from the origin; they just are not necessarily proportional to the distance from the origin. CRS production gives a cost function along the lines of $YC(w, r, 1)$, but homothetic production gives a cost function like $g(Y)C(w, r, 1)$. The equations also gave

$$\Delta Y = \epsilon^D \Delta P,$$

which just says we will move along the industry demand curve. Finally, we had the substitution equation

$$\Delta L - \Delta K = \sigma(\Delta r - \Delta w).$$

In the two-factor case, or with a homothetic production function, the relative factor demands depend only on the relative price of those two inputs. There is also a generalization of the elasticity of substitution that is useful for the multiple factor case: the partial elasticity of substitution,

$$\sigma_{ij} = \left(\frac{\partial^2 C}{\partial w_i \, \partial w_j} C \right) \Big/ \left(\frac{\partial C}{\partial w_i} \frac{\partial C}{\partial w_j} \right),$$

where C refers to the cost function, and i and j refer to inputs i and j. In the three-factor case, for example, it now matters whether we increase the relative price for input i by increasing the price of input i or decreasing the price of input j.

Recall from our demand system analysis that second derivatives of the cost function (equivalently, price derivatives of the conditional demand functions) are restricted by adding up. In partial-elasticity-of-substitution format, adding up looks like:

$$0 = \sum_j s_j \sigma_{ij},$$

where the s's are factor-spending shares.

PROPERTIES OF THE MULTIPLE-FACTOR INDUSTRY MODEL

Remember that

$$P = \frac{\partial C}{\partial Y}$$
$$Y = D(P)$$
$$X_i = \frac{\partial C(w_1, \dots, w_N, Y)}{\partial w_i}$$
$$Y = F(X_1, \dots, X_N).$$

How do we think about factor demand in this world? Impose constant returns to scale and consider, at the industry level, the derivative $\frac{\partial X_i}{\partial w_j}$. Using the first equation, note that $\frac{\partial^2 C}{\partial Y \partial w_j} = \frac{dP}{dw_j} - \frac{\partial^2 C}{\partial Y^2} \frac{dY}{dw_j}$. The second term cancels, because under CRS, $\frac{\partial^2 C}{\partial Y^2} = 0$. Now we're left with $\frac{\partial^2 C}{\partial Y \partial w_j} = \frac{dP}{dw_j}$. But $\frac{\partial^2 C}{\partial Y \partial w_j} = \frac{\partial^2 C}{\partial w_j \partial Y}$, $\frac{\partial C}{\partial w_j} = X_j$, and $\frac{\partial X_j}{\partial Y} = \frac{X_j}{Y}$ by CRS. So, $\frac{dP}{dw_j} = \frac{X_j}{Y}$. This means that the change in price due to a change in a factor price is just how much of the factor the industry uses per unit of output. This is related to the fact that CRS implies marginal shares are equal to average shares. Using the second equation, derive that $\frac{dY}{dw_j} = \frac{\partial D}{\partial P} \frac{dP}{dw_j}$; use the third equation to get that $\frac{dX_i}{dw_j} = \frac{\partial^2 C}{\partial w_i \partial w_j} + \frac{\partial^2 C}{\partial w_i \partial Y} \frac{dY}{dw_j}$. Finally, plug in for the relevant terms to achieve[1]

$$\frac{dX_i}{dw_j} = \frac{\partial^2 C}{\partial w_i \partial w_j} + \frac{X_i}{Y} \frac{\partial D}{\partial P} \frac{X_j}{Y}.$$

As before, the left term on the right-hand side is the substitution effect, and the right term on the right-hand side is the scale effect. For $i \neq j$, in the two-input case, the substitution effect is positive. In the multiple input case, if w_j increases, the substitution effect must on average be positive for other inputs. Now note that we can rewrite this entire equation in terms of elasticities:[2]

$$\frac{w_j}{X_i} \frac{dX_i}{dw_j} = \left(X_j w_j \frac{\partial^2 C}{\partial w_i \partial w_j} C \right) \bigg/ \left(PY \frac{\partial C}{\partial w_i} \frac{\partial C}{\partial w_j} \right) + \frac{P}{Y} \frac{\partial D}{\partial P} \frac{X_j w_j}{PY}$$
$$\Leftrightarrow \epsilon_{ij} = s_j \sigma_{ij} + s_j \epsilon^D.$$

As before, $s_j \sigma_{ij}$ is the substitution effect and $s_j \epsilon^D$ is the scale effect. Note that we can decompose the own-price elasticity in this way as well:

$$\epsilon_{ii} = s_i \sigma_{ii} + s_i \epsilon^D = -\sum_{j \neq i} s_j \sigma_{ij} + s_i \epsilon^D.$$

So, own-price elasticity is a share-weighted average of substitution elasticities and the output elasticity. The second equality is adding up for the second derivatives of the cost function.

This returns us to our discussion of Marshall's law from earlier in the book. There is no clear relationship between s_i and own-price elasticity. This is because some s_j change when s_i changes (s's sum to 1). Marshall (1890) noted that if a firm produces an intermediate input, it can raise the price significantly without much change in quantity if that intermediate input were a small factor for customers. This only incorporates the scale effect, however. If the substitution effect is more important, this is backwards, as Hicks pointed out.[3] Recall that Marshall's (1890) other points were correct: if output demand ϵ^D is more elastic, then demand for the input is also more elastic. Further, own-elasticity of input i is negatively related to how substitutable i is with other inputs.

ANALYZING PRODUCTION

We can think about the production function $F(X_1, \ldots, X_N)$ directly, or the cost function $C(w_1, \ldots, w_N, Y)$. These yield different first-order conditions. The first problem gives

$$P \frac{\partial F}{\partial X_i} = w_i.$$

But the second problem gives

$$X_i = \frac{\partial C(w_1, \ldots, w_N, Y)}{\partial w_i}$$

$$P = \frac{\partial C(w_1, \ldots, w_N, Y)}{\partial Y}.$$

For different problems, it will be useful to use different methods. If we want to hold quantities of other inputs constant, it is convenient to use the production function directly. If we want to hold prices of other inputs constant, it is convenient to use the cost function. In practice, what we hold constant corresponds to different experiments.

ENDOGENOUS FACTOR PRICES

What we just did assumes that when we change w_i, all other prices remain constant. Consider the following model. We have a supply function of X_j, $X_j^S(w_j)$, which is upward sloping. Now we have the equilibrium conditions

$$
\begin{aligned}
X_j^S(w_j) &= X_j(w_1, \ldots, w_N, Y) \\
X_i &= X_i(w_1, \ldots, w_N, Y) \\
P &= \frac{\partial C(w_1, \ldots, w_N, Y)}{\partial Y} \\
Y &= D(P).
\end{aligned}
$$

Now, w_j is endogenous. As we change the price of factor i, we now allow w_j to change. For a purely short-run analysis, make the elasticity of supply 0, so that $X_j^S(w_j)$ is fixed.

Just like in the consumer problem, there are multiple ways to think about the world. In the consumer problem, we had the Marshallian approach and the Hicksian approach. One is easy for some questions, and the other for different questions. We can use the same kind of logic here. For this problem, it would be easier to use the production function approach, because taking partial derivatives of the cost function holds prices constant.

Homework Problems for Part II

Market Equilibrium

1) Uber is a business that matches automobile drivers with paying passengers using an electronic mapping system. When a passenger enters Uber's market using his smartphone, he sees the price p and his approximate waiting time t for a vehicle/driver. For simplicity, assume that the price is quoted per minute of driving time. The revenue paid by each passenger is delivered to the driver, minus a 25% UBER handling fee.

   a. Does it matter whether the Uber handling fee is paid by passengers or drivers?

   b. If Uber ran an experiment of raising the posted price for a period of time, what would happen to passenger wait times and the number of rides? Would it matter whether Uber had announced (to potential drivers and passengers) this experiment in advance?

   c. What can you say about Uber's optimal price p and percentage handling fee?

   d. Could regulators use a price ceiling to encourage more efficient outcomes?

2) It is often alleged that the "war on drugs" leads to high levels of profits for the "companies" engaged in the international drug-smuggling business. Consider the case where competing suppliers of drugs to the U.S. market have access to a perfectly elastic supply of drugs overseas and import these drugs into the United States. Assume that these smugglers are heterogeneous with rising marginal costs of production

(here "production" means sneaking drugs into the country and selling them) and sell drugs in the United States in a competitive drug market.

a. Absent any enforcement by the U.S. government, what would determine the price of drugs, drug consumption, and the profits of the drug smugglers? How would changes in the price of drugs overseas affect the price of drugs in the United States?

b. Now assume that the U.S. government engages in an interdiction program by which it captures and destroys a fraction, f, of the drugs brought into the United States. How would this interdiction effort affect drug prices, drug consumption, and the profits of drug smugglers?

c. Now assume that rather than being intercepted by the U.S. government, the fraction, f, of shipments intercepted is captured by other criminals who then sell the drugs in the U.S. market. How will a change in f affect prices, consumption, and profits in this case? For an equal level of f, how will prices, consumption, and profits compare in this case to the case in part (b)?

3) An industry has a (large) fixed number of identical firms. Each firm produces the single output Y, with labor and capital according to the production function, $Y = F(L, K)$. The industry as a whole faces a downward-sloping demand for its output and an upward-sloping supply of labor. Capital is perfectly elastically supplied to the industry and all markets are competitive.

a. How will the elasticity of supply for an individual firm compare to the elasticity of supply for the industry as a whole?

b. How will the elasticity of demand for labor for an individual firm compare to the elasticity of demand for labor for the industry as a whole?

c. How will the effect of a change in the price of capital on output differ for an individual firm and the industry as a whole?

d. How will the effect of a change in the price of capital on the quantity of labor utilized differ for an individual firm and the industry as a whole?

4) Farmers on Long Island, New York, traditionally shipped their products to Cuba. *True, False, or Uncertain*: Those farmers were harmed by the U.S. government's embargo prohibiting exports to Cuba.

5) An overnight hail storm seriously dented the cars on display at the new-car dealership. Customers were aware of the natural destruction. *True, False, or Uncertain*: Therefore, fewer people shopped at the dealership until such time that it could obtain a fresh shipment of new cars.

6) *True, False, or Uncertain*: A reduction in the productivity of Illinois farm land would increase the number of acres in the state used for farming, especially to the degree that households' demand for food is price inelastic.

Part III

Technological Progress and Markets for Durable Goods

Chapter 15

Durable Production Factors

When we think about production, we can think about three types of inputs. There are inputs like labor, which are purchased in a service market; inputs like capital, which are purchased in a capital goods market; and materials, which are "used up" in the production process. Materials are not like capital assets because they do not have durability. Capital yields a flow of services over an extended period of time.

Land is a standard example of capital. It was, in fact, the most important form of capital for much of history. Initially, however, there was not a notion of "investment" in land. In more recent times, we have created land, investing on the quantity side. More frequently, however, we invest in the quality aspect of land. We make land usable by draining swamps, eradicating diseases, or terracing mountains.

A home is also a good example of capital. It provides shelter, privacy, storage, etc., year after year.

So there is a quantity side of capital. $K_t = K_{t-1} + I_t - DEP$, where I denotes investment and DEP denotes depreciation. Examples of this type of investment include building new houses and draining a swamp to create new land. Most commonly in economics, we let $DEP = \delta K_{t-1}$. That is, a constant fraction of capital yesterday depreciates in every period. We also do not care whether the capital from yesterday was built last year or 10 years ago. It all depreciates at the same rate:

$$K_t = I_t + (1-\delta)I_{t-1} + (1-\delta)^2 I_{t-2} + \ldots = I_t + (1-\delta)K_{t-1}.$$

The entire investment history is summarized in K_{t-1}. Exponential depreciation embodies the idea that capital of various ages is perfectly substitutable (just not at one-for-one rates) and also the idea that capital depreciates at the same rate regardless of how old it is.

Now that we are considering durable goods, we need to consider both the stock of durable assets and the flow of new investment. When the assets are very durable, these numbers can differ significantly. Consider the housing market: the stock of housing is much larger than the number of new houses built per year. The American Community Survey estimates that there are about 75 million single-family homes in the United States, whereas about 1 million single-family homes are built each year (U.S. Census Bureau 2017a,b).

There are also two notions of price, which we have mentioned before. There is the rental price, R_t, which is what we pay for capital for a moment in time, and a capital price, P_t, which is what we pay to purchase the asset and own the right to its use for the rest of its life. Note that this does not refer to renters versus buyers in the housing market, for example. Even if a house is purchased, the owner is still paying the rental price in each period because the owner could be renting the house and charging a tenant the rental price. Just as we could write K_t in terms of last period's capital *or* the perpetual inventory method, we can write

$$R_t = P_t - \frac{P_{t+1}(1-\delta)}{1+r} = \frac{rP_t + \delta P_{t+1} + P_t - P_{t+1}}{1+r}.$$

So, the rental price is the interest cost plus the depreciation cost plus the capital loss (a negative number if price goes up), expressed in next year's dollars. Or we could rearrange to put tomorrow's purchase price on the left, which would yield an estimate of the expected price tomorrow as a function of today's rental and purchase prices.

Yet another way to look at it: the relation between rental price and capital gains is a difference equation that can be solved for the initial purchase price:

$$P_t = R_t + \frac{1-\delta}{1+r} R_{t+1} + \left(\frac{1-\delta}{1+r}\right)^2 R_{t+2} + \cdots$$

So the amount K of capital we have is a backward-looking concept, but the capital price P of an asset is a forward-looking concept.

THE USE AND INVESTMENT MARKETS
FOR CAPITAL GOODS

A simple model involves thinking about a use market. In the use market, there is a demand schedule for capital. At any point in time, there is a capital stock K_t. Because we assume the use market clears, the capital stock, together with the demand for capital today, determines the rental price R_t, as shown in Figure 15-1. We can say all of this without thinking about the supply side at all.

Note this is the *use* market, so we're talking about demand for use of housing, automobiles, durable manufacturing inputs, and so on. Now we can think about the supply side, which will deal with the production of capital. There are many models we could use for this, but we will use a simple model that assumes a supply curve $I(P)$, as shown in Figure 15-2. Suppliers of capital care about the price, P_t, not the rental price, R_t, because producing a unit of capital will return P_t over its life.

FOUR EQUILIBRIUM CONDITIONS

We have assembled four equations for each point in time.

1. *Rental market equilibrium*: $K_t = D(R_t)$. This can be explicit, if transactions are rental transactions (i.e., apartments), or implicit if purchases are made, because the owner still pays the rental cost every period by *not* renting out the capital.
2. *Purchase-price equation*: $P_t = R_t + \dfrac{1-\delta}{1+r} R_{t+1} + \dfrac{(1-\delta)^2}{(1+r)^2} R_{t+2} + \cdots$. This is the forward-looking part of the model.
3. *Investment-good market equilibrium*: $I_t = I(P_t)$. This is the simplest possible supply model for investment.
4. *Law of motion for capital*: $K_t = (1-\delta)K_{t-1} + I_t$. The stock equation is the backward-looking equation. Knowing the capital we had yesterday is important for knowing the capital we have today.

The four unknowns at each point in time are the two prices and the two quantities.

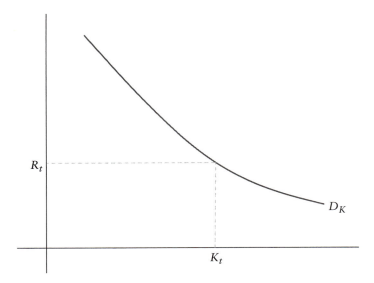

Figure 15-1: The use market. In equilibrium, the quantity of capital demanded today, K_t, along with its associated price, R_t, lie on the demand curve.

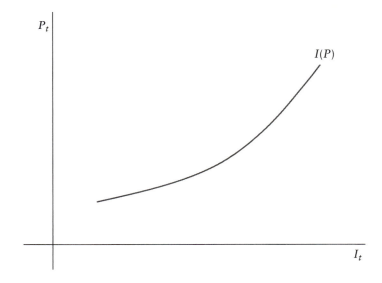

Figure 15-2: Investment market. Investment today, I_t, is chosen based on the capital price P_t.

We could, for example, endogenize δ. Consider a simple model of maintenance, where one can choose an optimal value of δ given a cost for reducing it:

$$\max_{\delta} P_{t+1} \frac{1-\delta}{1+r} - c(\delta).$$

This is an alternate form of investment. Expanding our conception of the model a bit—note that to get more capital tomorrow, one can produce more capital in addition to slowing depreciation by investing more into maintenance. For now, maintenance has simply been absorbed into the investment supply function $I(P)$.

STEADY STATE

Let's write these four equations for the steady state, where an upper bar denotes steady state:

1. $\bar{K} = D(\bar{R})$. Capital in steady state must equal the amount of capital demanded at the steady-state rental rate.

2. $\bar{P} = \dfrac{\bar{R}(1+r)}{r+\delta}$, just by condensing the geometric series.

3. $\bar{I} = I(\bar{P})$

4. $\bar{K} = \dfrac{\bar{I}}{\delta}$

We can use equation 1 and combine equations 2, 3, and 4 to write down steady-state supply and demand equations that pin down the steady-state rental rate and steady-state capital. The corresponding supply and demand curves are shown in Figure 15-3.

PERTURBING THE STEADY STATE

But if all we think about is the steady state, the concept is not very interesting. More interesting is the fact that the steady state tells us about where the system is headed. If changes put the system on a path toward a new steady state, what does that path look like? Suppose we were in a steady state, but a hurricane suddenly destroys some capital. Now we have some capital $K_0 < \bar{K}$. Then rents must be higher today. But then the

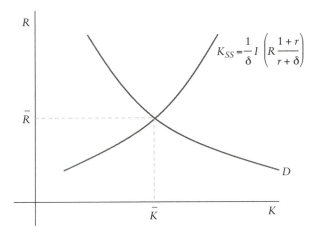

Figure 15-3: Steady-state rental rate and capital.

price also suddenly rises today, because the price is just the present value of rents. So investment also increases today, which will lead K to begin to rebound. These dynamics will continue working, and we will asymptotically return to the steady state. See Figure 15-4.

Now, think about how one would solve this system. Given K_0, R_0 is also known. But what would P_0 be? Think about guessing a value and call it $\hat{P}_0$. If $\hat{P}_0$ is too high, investment will initially rise a lot, driving the capital stock up quickly, which will lower rents too quickly to justify the high value for $\hat{P}_0$. If $\hat{P}_0$ is too low, we have the opposite problem. Investment will rise only slightly, which means the capital stock will not rise much. Then rents will stay high, and these high rents will imply that $\hat{P}_0$ should have been higher.

The immediate jump from $\bar{P}$ to P_0 is the efficient markets hypothesis. We are assuming that all the information about the new future is incorporated into the price today. The jump from $\bar{P}$ to P_0 is the capitalization of the information that became available.

What if people have naïve forecasts? For example, suppose people think the higher rent will stay high forever. Then P would rise more than P_0, and I would jump above I_0. But then R would fall quickly, and P would fall quickly, so we would get much faster convergence.

Now we will consider a rise in demand. We will move to a new steady state with a new level of output and price, as shown in Figure 15-5. When demand rises, we know capital in the future will be higher, as will rents. The figure shows us this. Further, we know that the price will be higher,

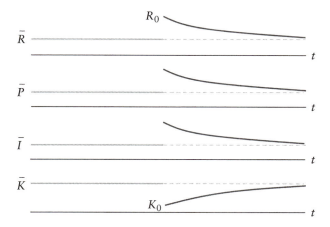

Figure 15-4: The system returning back to steady state. An initial negative shock to capital, which falls from $\bar{K}$ to K_0, causes rents to rise to R_0. Rents are related to the price, however, so higher rents in the future cause the price to rise today. Since the price is higher today, investment also jumps today.

as rents are higher, and we know that investment has to have risen as well. Depending on the elasticity of supply, the effects of higher demand will show up more in capital or in the rental price. If supply is very inelastic, most of the higher demand will appear in the new rental price, whereas elastic supply results in most of the higher demand will appear in the new steady-state level of capital.

The dynamics of an increase in demand are depicted in Figure 15-6. Note that the dynamics here are very similar to when we destroyed capital. While the new steady states are higher, unlike before, we are still engaged in a process of building new capital to meet demand.

Note that pinning down the higher price that results from higher rent involves firms making good forecasts about the future. If the pinned-down price is too high, investment will be too high, the stock of capital will rise quickly, and rents will fall quickly. Similarly, if the pinned-down price is too low, investment will be too low, the stock will rise slowly, and rents will fall only gradually. This problem of pinning down capital becomes very difficult when the capital is long-lived and discount rates are low—forecasts must be accurate far into the future.

Consider the case of nondurable goods in this model; that is, consider capital that lasts for only one period. If demand jumps, then all variables move right to the steady state immediately. The difference for durable goods is that rents, prices, and investment will all jump more than they

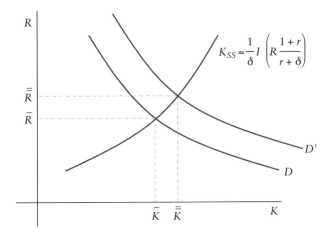

Figure 15-5: A rise in demand. The day demand jumps, rents will rise sharply, to the intersection of $\bar{K}$ with the new demand curve, D'. This results in the price rising sharply, causing investment to rise, and further causing capital to begin to move to the right, from $\bar{K}$ to $\bar{\bar{K}}$.

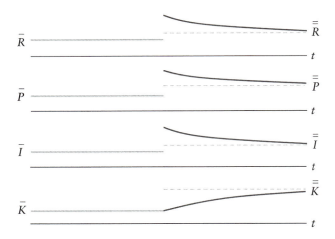

Figure 15-6: Transition dynamics for a rise in demand. Since capital is fixed in the short run, rents rise sharply. So prices rise sharply today, causing investment to rise drastically. This causes capital to begin to converge to the new steady state. As the capital stock converges, rents fall, causing prices to fall. Investment slows over time, and we converge to a new steady state.

would in the nondurable case. And they jump more the more durable the asset is. Consider a very long-lived asset with $\delta = .02$. This asset will survive for 50 years. In steady state, we are only producing 2% of the stock per year. If demand increases suddenly by 10%, investment will rise dramatically. This is one reason why investment is so much more

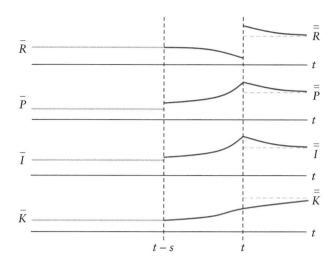

Figure 15-7: At time $t-s$, information about higher demand in the future is announced. Prices and investment jump today, causing rents to fall and capital to rise. When the higher demand is realized at period t, rents jump.

volatile than consumption. Despite the fact that demand rises in a given period, the stock has not moved significantly, so consumption cannot move significantly. Investment, on the other hand, is free to adjust much more in a short period. Suppose, even further, that a durable asset A' is used to produce a durable asset A. If demand for A rises, and investment jumps significantly, investment in durable asset A' will have to jump even more to meet the significantly higher demand resulting from increased production of A.

Now suppose we have the same demand increase, but instead of happening today, we learn that the demand increase will happen in the future. Rents will not change today, but because rents will be higher in the future, the price rises today. As a result, investment increases today, and the capital stock begins to increase. Thus, rents begin to fall. Though rents are falling, prices continue to rise because the higher demand in the future is less discounted. Thus, in the short run, rental rates will fall. See Figure 15-7 for a depiction of these dynamics.

Note that we cannot build up all of the required capital for the new steady state in advance. Suppose we did—we will derive a contradiction. If $K_t = \overline{\overline{K}}$, then it must be that $R_t = \overline{\overline{R}}$ and that R continues to equal $\overline{\overline{R}}$ after period t. Prior to period t, R must be below $\overline{R} < \overline{\overline{R}}$ because demand has not yet increased and the stock is above $\overline{K}$. But then we

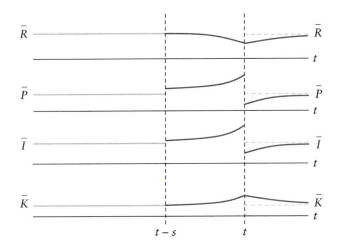

Figure 15-8: At date $t-s$, we learn that demand might rise on date t. At date t, we learn that guess was wrong.

must have that prices before period t were less than $\overline{\overline{P}}$, which implies investment before period t must be less than $\overline{\overline{I}}$. If we are investing less than $\overline{\overline{I}}$ prior to period t, however, there is no way we could have reached $\overline{\overline{K}}$.

These dynamics reflect the ideas of rational expectations and market efficiency. We expect that upon learning new information, the prices today jump to reflect this new information. This is why we see change in period $t-s$; prices today are changing to reflect the new information. These dynamics reflect boom periods we see in the real economy. In the housing market, for instance, when people think demand is going to be high in the future, rents are low but prices are high and rising. This process describes a fully rational housing boom. An observation of low rents together with high and rising prices is no proof of irrational pricing.

Now let's suppose that at time $t-s$, we learn that there is a chance that demand will increase at date t. At date t, we actually learn that this hunch was wrong; demand remains unchanged. These dynamics are depicted in Figure 15-8. Now, when we learn demand has remained unchanged, we have too much stock. Instead of jumping, rents remain low. Prices fall drastically, and as a result investment falls as well. Over time, the capital stock reverts to the steady state, causing rents to rise, prices to rise, and investments to rise back to the previous steady-state level.

Now, this story looks a lot like the housing bust circa 2007. How do we tell the difference between this story and the "irrational exuberance" story? We need to think about whether or not it was reasonable to think demand was going to be high in the future. We can always write the following equation:

$$P_t = R_t + \frac{R_{t+1}(1-\delta)}{1+r} + \frac{R_{t+2}(1-\delta)^2}{(1+r)^2} + \cdots + \frac{R_{t+N-1}(1-\delta)^{N-1}}{(1+r)^{N-1}} + \frac{P_{t+N}(1-\delta)^N}{(1+r)^N}.$$

Today's price must be justified by the rents we expect to receive in the future, plus some terminal value. In the housing boom, we know that rents were low, but prices today were high. This means people thought prices were going to be high in the future. It's easy to say after the fact that what we saw in the housing boom was just an irrational bubble. But before the crash happened, if we say that people had expectations that future prices were going to be high; that is, there was some high terminal value, determining whether the expectations were rational after the fact is very difficult. Put differently, in the housing crisis, it turned out that people's expectations were wrong relative to what happened. The question remains, however: were people wrong about what *could* have happened?

Chapter 16

Capital Accumulation in Continuous Time

PERTURBING THE STEADY STATE (CONTINUED)

In the previous chapter, we analyzed a model of capital investment and applied it to the housing market. There were several major points here. During the boom period, there were two key features: prices were high, so producers found it an attractive time to produce homes, but rents were low, so it was cheap (on net) to live in homes. Even purchasing an expensive home was "cheap" because people believed homes were appreciating in value.

During the bust period, after the housing market crashed, prices plummeted. In fact, they often fell below construction cost. It looked like buying houses was profitable during the bust. The problem was that it is hard to police depreciation, especially when renting out single-family homes. We can start to formalize some of these ideas.

Consider what occurs in the market when the interest rate is reduced. Because interest rates appear in the denominator for the present value of future rents, then for given rent values, a lower interest rate means a higher price. This shifts the steady-state supply curve to the right. In the new equilibrium then, $\bar{\bar{R}} < \bar{R}$, $\bar{\bar{I}} > \bar{I}$, $\bar{\bar{K}} > \bar{K}$, and $\bar{\bar{P}} > \bar{P}$. This looks much like a boom period. This change is depicted in Figure 16-1, and the transitional dynamics are shown in Figure 16-2. During the housing boom, note that we also had decreasing rates, which contributes even more to the housing boom effects we see.

Now let's think about the speed of convergence to the steady state. Several parameters can be important here. The depreciation rate, for example, is very important. Suppose I, for example, is constant. Then

166

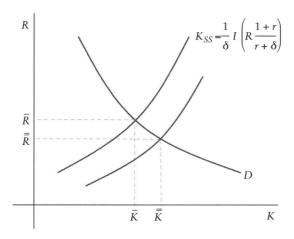

Figure 16-1: A decrease in the interest rate shifts the supply curve to the right. In the new steady state, there is more capital, lower rents, higher prices, and higher investment.

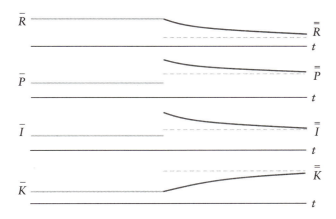

Figure 16-2: These transitional dynamics result from a decrease in the interest rate. Price increases immediately, driving up investment, which causes rents to begin to fall and the capital stock to begin to rise.

the convergence rate is simply given by δ. In the continuous case, $\dot{K} = -\delta K + I$, where $\dot{K}$ gives the instantaneous rate of change of capital. This equation shows that convergence under these assumptions will occur at rate δ. Also, the elasticity of supply is important. A higher elasticity means faster convergence. Finally, the demand elasticity has an effect, since less elastic demand implies faster convergence. When the stock is reduced, for example, rents jump up more when demand is more inelastic. This causes prices to jump higher and investment to rise higher, which implies that the capital stock will rise faster to meet the new demand.

Note that learning about a demand change, for example, far in advance of it occurring does not mean that the stock will reach the new steady state in advance of the demand change occurring. Building homes prior to the demand increase means losing some money in the short run because the new houses are not yet demanded. Similarly, however, building houses in excess of steady-state investment after the demand increase means that more money could be made if those houses existed already. The costs of building houses early, however, prevent the new steady state from being achieved before the higher demand is realized.

The future is almost as important as the past when the interest rate is low. The past is much more important than the future, however, when interest rates are high.

CONTINUOUS-TIME VERSIONS OF THE FOUR EQUILIBRIUM CONDITIONS

Now, let's think about this entire model in continuous time. The equations become

1. Rental market equilibrium: $K(t) = D(R(t))$.
2. Purchase-price equation: $P(t) = \int_t^\infty e^{-(r+\delta)(\tau-t)} R(\tau) d\tau$.
3. Investment-goods market equilibrium: $I(t) = I(P(t))$
4. Law of motion for capital: $\dot{K}(t) = -\delta K(t) + I(t)$.
2′. $R(t) = (r+\delta)P(t) - \dot{P}(t)$

Sometimes equation 2′ is considered a primitive of the model. But equation 2 holds for all t, so one can differentiate it with respect to t. Equation 2 implies equation 2′. Note, importantly, that equation 2′ does not imply equation 2. Equation 2′ does not include the boundary condition that says prices must be set at the correct level. This is important, because we will now use equations 1, 3, 4, and 2′ in Figure 16-3.

We can combine equations 1 and 2′ to form Equation 5: $K(t) = D((r+\delta)P(t) - \dot{P}(t))$. We can further combine 3 and 4 to form Equation 6: $\dot{K}(t) = -\delta K(t) + I(P(t))$. Now everything is in terms of K, P, and

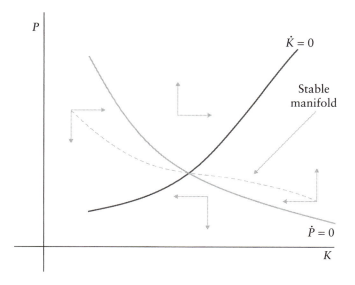

Figure 16-3: A phase diagram. The lines $\dot{K}=0$ and $\dot{P}=0$ denote when capital and price are constant, respectively. The stable manifold is the line along which an economy converges to the steady state, where $\dot{P}=\dot{K}=0$.

their derivatives. We will use these equations to draw a phase diagram. It is useful to consider the steady state when $\dot{K}(t)=0$ and $\dot{P}(t)=0$.

Above the $\dot{K}=0$ line, the price is high, so the capital stock is increasing. Above $\dot{K}=0$ on Figure 16-3, we therefore draw arrows pointing to the right. Similarly, below $\dot{K}=0$ on Figure 16-3, we denote by a left arrow the fact that the capital stock is shrinking over time because prices are too low. To the right of the $\dot{P}=0$ line, the level of capital stock is too high, so the price is too low and therefore rising over time. We denote this with an up arrow. The opposite is true when we are to the left of the $\dot{P}=0$ line, so we denote the movement of price to the left of this line with a down arrow. All of these dynamics can be easily shown mathematically using equations 5 and 6.

Convergence to the steady state happens along the saddle path or stable manifold, denoted by a dashed line on the phase diagram in Figure 16-3. As an example, suppose demand unexpectedly rises today, and Figure 16-3 shows the new steady state in the world with higher demand. Then the capital stock, as we know, is too low today to meet the higher demand. Rents rise, causing prices to jump today. Prices cannot just jump simply an arbitrary amount, however. The saddle path

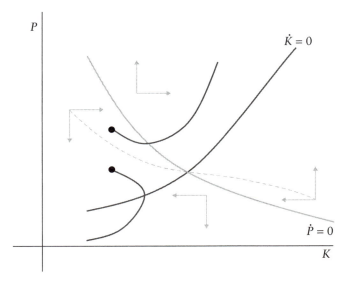

Figure 16-4: If price jumps to the wrong point, the dynamics do not lead the economy to converge to the steady state.

gives the unique value to which the old price must jump in the presence of higher demand.

What happens if price jumps above or below the saddle path? This is shown in Figure 16-4.

Finally, consider a myopic model in which we replace the rational expectations equation for $P(t)$ with

$$P(t) = \frac{R(t)}{r+\delta}.$$

Then future anticipated shocks would have no effect. Further, we would adjust to the new steady state faster, since instead of jumping to the saddle path, P jumps to the demand curve. Note that the approached steady state will be the same, and that this will not look that different from the rational expectations case. Finally, note that under this model, holding capital is a bad deal if prices are falling, since the cost is actually higher than $R(t)$.

Lastly, observe that if investment cannot be negative, then uncertainty causes people to underinvest. Suppose there's a 50% chance of a 10% increase in demand and a 50% chance of a 10% decrease. Then the initial jumps in rental rate will be equal (in opposite directions), but in the case of the negative shock, the rent will return to steady state much more slowly because the stock cannot adjust as quickly downward given that there is no negative investment. So on average rents are lower, and thus the price today must be lower.

In the next chapter, we look at a model that has an endogenous interest rate instead of a simple upward-sloping supply curve, as we have been assuming.

Chapter 17

Investment from a Planning Perspective

Last time, we considered a very simple investment model. The key underlying fact was that we had an upward-sloping supply of investment. That's why the dynamics we discussed came about. It is cheaper to build the product more slowly than it is to build it all at once. There are a number of reasons for this in practice: heterogeneity of underlying resources, for example, or multiple capital goods required as factors of production. With upward-sloping supply, production is smoothed over time to reduce the total cost of production. This is like a convexity in the cost structure. The upward-sloping supply curve, or a rising marginal cost, corresponds to convexity in overall cost. As a result, the behavior we had previously represented as market equilibrium involving multiple agents (e.g., renters of capital, builders of new capital goods) can be restated as a maximization problem of a single producer.

Take the inverse demand function $D(x)$ and define the function $V(x) = \int_0^x D(z)dz$. Then, by the fundamental theorem of calculus, $\dfrac{dV(x)}{dx} = D(x)$. Similarly, take the supply function $S(x)$ and define $C(x) = \int_0^x s(z)dz$; then, by the same reasoning, $\dfrac{dC(x)}{dx} = S(x)$. Then consider the problem of maximizing total surplus:

$$\max \int_0^\infty e^{-rt}[V(K(t)) - C(I(t))]dt.$$

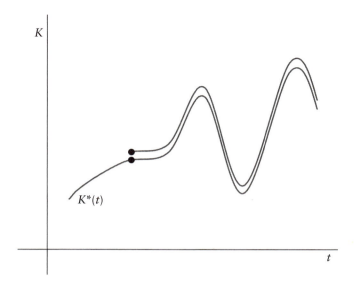

Figure 17-1: Increasing $\dot{K}$ for one period and maintaining that additional capital over the equilibrium path, $K^*(t)$, for the remaining periods.

Because of how we have defined V and C, taking derivatives here gives the typical first-order conditions in terms of $\dfrac{dV(x)}{dx}$ and $\dfrac{dC(x)}{dx}$. But these are just demand and supply. In other words, we have represented an equilibrium model where quantity supplied equals quantity demanded as simply a maximization problem. Note that this is not maximizing the true social surplus, unless we think that the demand curve represents marginal social value and that the supply curve represents marginal social cost. The true equilibrium, however, is generated by taking derivatives, since we get back to the demand and supply curves. To find the equilibrium, we can rewrite this problem as

$$\max \int_0^\infty e^{-rt}[V(K(t)) - C(\delta K(t) + \dot{K}(t))]dt,$$

because $\dot{K} = -\delta K + I$. Note that, when we take the derivative here, we are not choosing both K and $\dot{K}$. If we know the initial level of capital, for example, as well as $\dot{K}$ for every period, then we also know K in every period. Consider the experiment where we take a proposed path $K^*(t)$ for capital and from there consider changing $\dot{K}$ for one period and then maintaining that additional capital in future periods—that is, we change $\dot{K}$ in one period but fix $\dot{K}$ in future periods, as illustrated in Figure 17-1.

In this experiment, we are maintaining the same increment to capital after date t, which means that gross investment has to be incremented enough. If we have capital on an optimal path, then this experiment—for that matter, any deviation from the path—cannot add to our objective. In other words, the derivative of the objective is zero:

$$C'(t) = \int_t^\infty e^{-r(\tau-t)}[V'(\tau) - \delta C'(\tau)]d\tau$$

where we have condensed the notation and replaced each function's argument with just the time index for the argument.[1]

This is just the fact that marginal cost of investing must be the same as the present value, discounted at rate r, of having a bit more capital at all future dates. We typically refer to $[V'(\tau) - \delta C'(\tau)]$ as the net return on capital, so the right-hand side is the present discounted value of the net return on capital.

Alternatively, we could consider investing in one more unit of capital in a given period but then simply letting it depreciate. This means changing $\dot{K}$ for one period and then holding I fixed at all other points in time. This experiment also cannot add to the objective:

$$MC = C'(t) = \int_t^\infty e^{-(r+\delta)(\tau-t)}V'(\tau)d\tau$$

where we are now discounting at rate $r + \delta$ because this experiment involves adding smaller bits to capital at dates more distant in the future. An optimal path not only involves marginal cost equal to the (r-discounted) present value of the future net returns on capital, but it also involves marginal cost equal to the ($r + \delta$-discounted) present value of the future gross returns on capital.

Both of these approaches are equally valid for modeling the world, but be careful to be consistent! Many people often confuse gross and net returns when working with these problems.

ADJUSTMENT COSTS APPLIED TO NET INVESTMENT

Prior to this, we have considered the approach using *gross* returns because we set marginal cost equal to the present value of rental payments discounted at $r + \delta$. Now, let's eliminate the assumption that the supply price of investment is rising—that is, before we had a rising short-run and long-run supply of capital. Let's consider an adjustment cost model. This can be combined with upward-sloping supply, but for now let's leave supply perfectly elastic for simplicity. In other words,

$$C(I) = PI$$

so the marginal cost of investment is P. In this world, our old model would give us a highly volatile capital stock. Here, however, we will impose a cost to changing the capital stock, $A(\dot{K}) = \frac{1}{2}A\dot{K}^2$. It is therefore costly to increase or to decrease capital. Convexity gives us that faster stock adjustments are more costly. Note that this may not fit reality well in many circumstances because it will give that many tiny adjustments over time are preferred to few large adjustments. One could consider incorporating a fixed cost of adjustment, but this does not make much sense in a continuous time model. Certainly, this assumption is often acceptable for macroeconomic situations, which aggregate micro-level facts in a way that creates a smooth result largely consistent with our continuous time analysis. As we will see, however, the long-run behavior of this model is very different than the one we just analyzed. Now, we solve

$$\max \int_0^\infty e^{-rt}\left[V(K(t)) - P(\delta K(t) + \dot{K}(t)) - \frac{A}{2}(\dot{K}(t))^2 \right]dt.$$

As before, in Figure 17-1, consider perturbing $\dot{K}$ for one period and then holding it fixed in future periods. The first-order condition, differentiating with respect to $\dot{K}$ at date t, will be

$$-e^{-rt}(P + A\dot{K}(t)) + \int_t^\infty e^{-rt}(V'(\tau) - \delta P)d\tau = 0.$$

So,

$$MC = P + A\dot{K}(t) = \int_t^\infty e^{-r(\tau - t)}(V'(\tau) - \delta P)d\tau.$$

Thus, investment is not immediate in this model because adjustment of the capital stock is costly. But now consider the steady state:

$$P = \frac{1}{r}(V' - \delta P),$$

which implies

$$V' = rP + \delta P = (r + \delta)P$$

So in this model, though we have upward-sloping supply of capital in the short run, there is perfectly elastic supply of capital in the long run. The structural difference from before is that, while we had a rising supply of gross investment, now we only have a rising supply of net investment. That is, the adjustment cost function is a function of $\dot{K}$, not $\dot{K} + \delta K$.

ENDOGENOUS INTEREST RATES: THE NEOCLASSICAL GROWTH MODEL

Now, we will consider a third model. We will have a single good economy, where that good can be consumed or invested:

$$Y = C + I$$
$$Y = F(K).$$

So Y is produced using K, and Y is either consumed or invested. This formulation also means that C and I are perfect substitutes. There is no rising supply curve or adjustment costs here. The production possibility frontier is shown in Figure 17-2.

Now, the cost of increasing investment will be sacrificing consumption. Since we want to smooth consumption over time, we will in effect have a rising supply price. The consumer will solve

$$\max \int_0^\infty e^{-\rho t} U(C(t)) dt$$

where $C(t) = F(K(t)) - I(t)$ and ρ is the rate of time preference. Thus, $C(t) = F(K(t)) - \delta K(t) - \dot{K}(t)$. Rewriting the problem, we have

$$\max \int_0^\infty e^{-\rho t} U(F(K(t)) - \delta K(t) - \dot{K}(t)) dt.$$

Holding future $\dot{K}$ constant and increasing it today, we get the first-order condition

$$-e^{-\rho t} U'(t) + \int_t^\infty e^{-\rho \tau} U'(\tau)[F'(K(\tau)) - \delta] d\tau = 0$$

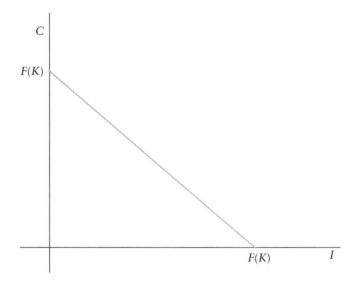

Figure 17-2: The production possibility frontier. *C* and *I* are perfect substitutes, but their combination must equal the total amount produced *F(K)*.

or

$$1 = \int_t^\infty e^{-\rho(\tau-t)} \frac{U'(\tau)}{U'(t)} [F'(K(\tau)) - \delta] \, d\tau = \int_t^\infty e^{-\int_t^\tau r(s)ds} [F'(K(\tau)) - \delta] d\tau.$$

The right-hand side is the market value of the future net returns on capital. The left-hand side is simply the number 1 because the cost of capital is 1 unit of consumption. This is a simple, consumption-based valuation of capital.

The dynamics will come from the fact that people do not want to starve themselves to build up capital quickly. Let's consider the steady state, where $U'(\tau) = U'(t)$:

$$\frac{F'(K) - \delta}{\rho} = 1,$$

which implies

$$F'(K) = \rho + \delta.$$

Thus, in steady state, the marginal product of capital only depends on two numbers, ρ and δ. Now suppose that in steady state, we change technology by doubling the output we get per unit of capital. That is, we

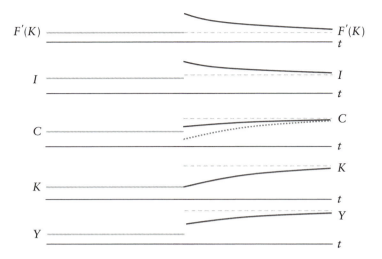

Figure 17-3: The effect of doubling F.

double F. The results of this experiment are depicted in Figure 17-3. The marginal product of capital will increase today, but we know it must remain the same in steady state, since ρ and δ are constant. Thus it falls toward the new steady state, which means capital is rising over time. Since F doubles, we also have that Y jumps. It pays to invest, because $F'(K)$ is above the steady state, so investment jumps. But we do not know what will happen to C in the short run. On the one hand, the agent is richer and will want to increase consumption so as to smooth it over time. On the other hand, it really pays to invest: because the marginal product of capital is so high, consuming today is expensive. Consumption may increase or decrease in the short run, but we know it will be higher in the long run. To see this, differentiate the first-order condition with respect to time to get

$$\dot{C}(t) = -\frac{U'(C(t))}{U''(C(t))}(F'(K(t)) - (\rho + \delta)).$$

Thus, because $U' > 0$, $U'' < 0$, and $F'(K) > \rho + \delta$, we get that $\dot{C}(t) > 0$. Whether consumption initially increases or decreases when the technology changes depends on the curvature of the utility function, but after the initial jump, it will be increasing toward the new, higher steady-state level. The two possible types of consumption paths are shown in Figure 17-3.

In either case, the investment path can slope down as shown in Figure 17-3, although with enough curvature in the utility function the investment path can slope up.

This model captures the essence of the neoclassical growth model. Because ρ is a constant, the long-run supply of capital curve is horizontal; we have perfectly elastic long-run supply. Otherwise, the supply curve could slope upward or downward depending on whether people become more patient or less patient as they become richer (i.e., that ρ depends on the level of consumption or income or capital).

What happens if the rate of time preference falls as people become richer? Now, long-run supply will actually be downward sloping. In the experiment depicted in Figure 17-3, if, as we become richer, we become more patient, the steady-state marginal product of capital will be below the original steady-state level as a result of doubling F.

This model is very much like the adjustment cost model. The adjustment cost here is the fact that adjusting capital quickly may mean drastically altering consumption. To see this another way, note that the interest rate in this model is endogenous. The interest rate r between t and τ is

$$e^{-r(\tau-t)} = e^{-\rho(\tau-t)} \frac{U'(\tau)}{U'(t)}.$$

The more one invests, the higher future consumption is relative to present consumption. Thus, the marginal utility from additional consumption in period τ is reduced relative to the marginal utility received today, driving the interest rate up. This dynamic prevents the economy from reaching the new equilibrium right away.

Chapter 18

Applied Factor Supply and Demand 1

Technological Progress and
Capital-Income Tax Incidence

DEFINITIONS OF LABOR PRODUCTIVITY

Now we want to think about technological progress. Most generally, we think about technological progress in terms of productivity. We measure this commonly through labor productivity,

$$\frac{Y}{L} = \text{Real Output per Worker Hour.}$$

We also have the measure

$$\frac{W}{P} = MP_L = \text{Marginal Product of Labor.}$$

These are the average and marginal products of labor, respectively. Finally, we can consider the ratio of these two numbers,

$$\frac{W/P}{Y/L} = \frac{WL}{PY} = S_L.$$

The share of labor stays roughly constant over time if and only if both measures of productivity grow roughly at the same rate.

EXPLAINING ECONOMIC GROWTH IN THE PRESENCE OF COMPLEMENTARITY

But what lies behind the growth in labor productivity and rising real wages? We have a few hypotheses here.

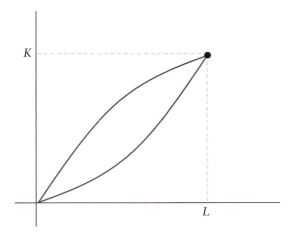

Figure 18-1: If I allocate along the margin according to the bottom path, there will be relatively abundant labor, and much of the output will be attributed to capital. If I allocate along the margin according to the top path, the opposite is true.

1. Capital deepening (K/L is rising).
2. Technological change (more output for the same inputs).
3. Human capital accumulation (perhaps people have more skills on average than they did before).
4. Better allocations (we use our resources better).

This fourth point tends to be considered on the macro level, whereas the first three can be considered on a micro level. Note further that it is difficult to consider these as independent explanatory factors. If I invent a new technology, for example, I am going to need capital to get my project going. Similarly, when we see $Y = F(K, L)$, we do not try to ask how much of the output came from capital versus labor. We do ask, however, how much each gets rewarded in the market.

In short, there's complementarity between all of these factors. All of the decompositions of the growth that we will do depend on the path we take to get there. For example, suppose to get more output we had to build a road and a vehicle. If the vehicle arrives after the road is already there, it seems like the vehicle is the key factor for getting the output. If the opposite occurs, however, the road looks like the key factor.

Take an output level $F(K, L)$. How we account for that output in terms of capital and labor depends on how we acquired those two factors. See Figure 18-1, which shows two paths. On the upper (lower) path, capital (labor) tended to accumulate first, respectively. Due to the complementarity in production, the marginal product of labor is high along the upper path, so we would attribute a lot of the additions to output to additions to labor. The opposite happens on the lower path.[1]

A cake, for example, can be changed by altering the amount of eggs, flour, or other ingredients on the margin. But in general, once the cake as a whole exists, we cannot decompose what "portion" of the cake comes from an individual ingredient.

THE CONSEQUENCES OF UNBIASED TECHNOLOGICAL CHANGE

Returning to our question of technological progress, for now we will focus on capital deepening and technological change. Note that $\Delta(Y/L) = \Delta Y - \Delta L$, where we are thinking about Δ as a percent change and make use of the log properties. The change in total factor productivity, TFP, will then be given by[2]

$$\Delta TFP = \Delta Y - \Delta Total\ Input$$
$$\Delta TFP = \Delta Y - (S_L \Delta L + S_K \Delta K).$$

Recall that in the constant returns to scale case, $S_L + S_K = 1$, since $F_K K + F_L L = F$, but CRS is not needed to formulate the general form of this equation given above. We can also measure TFP on the price side, just as we did with the real wage for labor productivity:

$$\Delta TFP = (S_L \Delta W + S_K \Delta R) - \Delta P.$$

This measure will include technological change, human capital change, and better allocations. Together we have that TFP growth reflects both output growing faster than inputs and factor prices growing faster than output prices.

For now, we are simply assuming growth is driven just by capital deepening and technological change. Then

$$\Delta Y - \Delta L = S_K(\Delta K - \Delta L) + \Delta TFP.$$

Thus, growth in labor productivity can be decomposed into capital deepening, $S_K(\Delta K - \Delta L)$, and technological progress, ΔTFP. These are both related, however. In the neoclassical growth model, technological growth from increasing the productivity of capital would induce capital deepening.

The price-based measure of TFP can also be rewritten:

$$\Delta TFP = S_L \Delta \frac{W}{P} + S_K \Delta \frac{R}{P}.$$

If both real wages and the real rental rate on capital are growing, then we must have technological progress.

What would the neoclassical growth model say if we have a large sudden improvement in technology, such as $\Delta TFP = \Delta \frac{W}{P} = \Delta \frac{R}{P} = 20\%$?

The economy will respond with capital deepening. Recall the elasticity of substitution:

$$\Delta \frac{W}{P} - \Delta \frac{R}{P} = \Delta W - \Delta R = \frac{1}{\sigma}(\Delta K - \Delta L).$$

In the short run, capital does not deepen (by definition) and the two real rental rates increase by the same percentage. As capital is accumulated, $\Delta(W/P)$ increases further and $\Delta(R/P)$ decreases. In the long run, we will have $\Delta \frac{W}{P} = \frac{\Delta TFP}{S_L}$. That is, all of the benefit of the TFP growth will accrue to the real wage. We also have that

$$\frac{S_L}{S_K} = \frac{WL}{RK} = \frac{\dfrac{W}{R}}{\dfrac{K}{L}},$$

but we don't know the way that this is moving, because both $\frac{W}{R}$ and $\frac{K}{L}$ are moving upwards.

We arrived at these long-run results with a supply and demand framework. On the factor-supply side, we said that capital is perfectly elastically supplied ($\Delta R = 0$), and labor is not ($\Delta W \neq 0$). From there, the price-based TFP measure immediately gives us that all of TFP goes to wages. When we add the factor demand curves—specifically, their differences in loglinear form—we additionally find that TFP increases the capital-labor ratio, with a magnitude that increases with the elasticity of substitution σ.

THE INCIDENCE OF A CAPITAL-INCOME TAX

We can augment this framework to address a number of other growth questions, but already it tells us about the incidence of a capital-income tax, which is a fraction $\tau \in (0, 1)$ of capital income RK that is paid to the government with the revenue distributed equally to all of the owners of labor. With such a tax in place, perfectly elastic capital supply means that $\Delta R + \Delta(1 - \tau) = 0$ because, by definition, no finite amount of capital is supplied unless the after-tax rental rate is the same as it would be without the tax. Let's hold TFP constant while we focus on the tax. From our price-based definition of TFP, we have that $0 = S_L \Delta W + S_K \Delta R$. This and the supply condition give us

$$\Delta W = \frac{S_K}{S_L} \Delta(1 - \tau) < 0$$

where we are considering a change from a low tax rate to a higher one. In other words, the capital-income tax reduces wages in the long run. The magnitude of this effect is decreasing in labor's share, which is interesting because labor's share has fallen over the past two decades. The long-run wage impact of capital taxation may be more negative than it used to be.

Bringing in the factor-demand equations, we have the amount that the tax reduces capital intensity, and therefore average labor productivity:

$$\Delta K - \Delta L = \sigma(\Delta W - \Delta R) = -\Delta R \sigma \left(\frac{S_K}{S_L} + 1 \right) = \sigma \frac{\Delta(1 - \tau)}{S_L} < 0$$

$$\Delta Y - \Delta L = S_K (\Delta K - \Delta L) = \frac{S_K}{S_L} \sigma \Delta (1 - \tau) < 0.$$

The workers are not only getting their wage, but also the revenue from the tax. Holding labor fixed, what we have above is enough to prove that workers lose more in wages than they gain in tax revenue from a marginal increase in the tax rate. As a share of aggregate income, the increase in workers' post-fisc income is:[3]

$$S_L \Delta(WL) + \tau S_K \Delta(\tau RK).$$

Recall that Δ denotes log changes: the first (second) Δ term is multiplied by labor income (tax revenue τS_K) so that it is an absolute change,

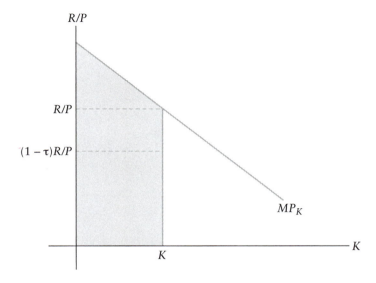

Figure 18-2: Output and factor incomes as areas under the marginal product schedule.

respectively. Using $\Delta L = 0$ and the results above for capital and wages, we have that workers' post-fisc income falls with the tax; the first term is more negative than the second term is positive (if at all).[4]

These results are easy to see in the capital demand diagram, Figure 18-2. For any given capital stock K, aggregate output (and therefore aggregate income) is the area under the demand curve to the left of that amount of capital.[5] The equilibrium capital-rental rate R/P vertically divides that area between labor income (above) and capital income (below), both before taxes and transfers. The capital income tax increases R/P and therefore reduces labor income, WL.[6]

To see the effect on post-fisc income, note that the pre-tax capital income rectangle can be further divided in two at the height of the after-tax equilibrium rental rate $(1 - \tau)R/P$, with tax payments above and after-tax capital income below. With perfectly elastic capital supply, the after-tax rental rate is independent of the tax rate, so that the tax must raise R/P and move the economy up the marginal product schedule, thereby reducing the sum of labor income and tax revenue. Even when labor receives all of the revenue, they still pay the deadweight loss of the tax: the extra output that additional capital would add beyond what it costs to supply that capital.

WHY CAPITAL IS ELASTICALLY SUPPLIED IN THE LONG RUN

The supply of capital reflects its marginal cost, which itself reflects (i) the willingness of people to delay consumption and (ii) the ability of producers to make investment goods rather than consumption goods. In the long run, consumption and investment are constant; whether they are constant at a high level or a low level depends on technology, taxes, and so on. But the level of consumption can affect (i) and (ii) in either direction. For example, people might be more patient when they are richer (higher consumption levels), which by definition means that capital is less costly to supply. In other words, this example has the supply of capital sloping down in its rental rate. On point (ii), the production of investment goods could be more capital intensive than consumption goods, which means that the economy with more capital in the long run supplies capital at a lower marginal cost in terms of forgone consumption.

The neoclassical growth model is neutral on points (i) and (ii). Its long-run capital-supply curve therefore neither slopes down nor up: it is horizontal. A horizontal long-run supply curve was the basis for this chapter's long-run analysis of the incidence of productivity and capital taxes.

THE INCIDENCE OF A CORPORATE-INCOME TAX

A corporate-income tax is different from a capital-income tax because the noncorporate capital does not pay corporate-income taxes. We address this by modifying the production function a bit so that it is $F(K, L)$, where K is now a homogeneous aggregate of two capital inputs: $G(K_1, K_2)$. This two-capital-input approach allows us to recognize that the effects of the corporate tax are different on corporate and noncorporate capital. Assume also that output can be used for consumption or either type of investment: $Y = C + I_1 + I_2$. Investment can go into either sector but the types of investment are perfect substitutes in terms of forgone consumption.

Because the aggregate is homogeneous, we can focus on the unit isoquants, which are the combinations of K_1 and K_2 that result in exactly one unit of aggregate capital K. See Figure 18-3. The cost of any one of those possibilities in terms of forgone consumption is where the line with slope -1 intersects the vertical axis (since the price of K_2 is 1 and not affected by the tax).

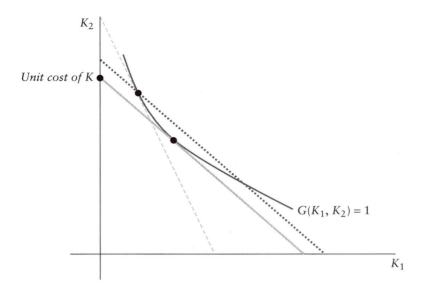

Figure 18-3: Using the unit isoquant to measure the cost of sectoral distortions.

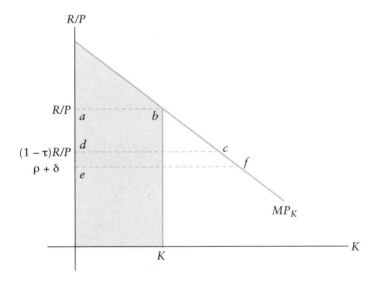

Figure 18-4: Labor's extra loss from distorting the composition of investment.

The social cost is minimized where the isoquant also has slope −1, as shown in the figure. But when K_1 is taxed more than K_2, investors will not minimize social cost because they want to avoid taxes too. Instead they have too little K_1. This raises the cost of each unit of K output by the vertical distance between the solid line and the dotted line.

Now let's go back to Figure 18-2. Previously, we said that the after-tax return on K is fixed, so that each unit of taxation increased the pre-tax rental rate R/P one-for-one. But now the after-tax return on K must also increase because it takes more investment to get each unit of K. See Figure 18-4. The loss to workers is not only the trapezoid *abcd* between R/P and $(1-\tau)R/P$, but also the trapezoid *cdef* between the new after-tax return and the old one.

In other words, holding tax revenue constant, labor loses more when that revenue is raised by taxing just some of the capital rather than all of it.

Chapter 19

Applied Factor Supply and Demand 2

Factor-Biased Technological Progress, Factor Shares, and the Malthusian Economy

THE DEFINITION OF TECHNOLOGICAL BIAS

The neoclassical growth model can be summarized as: in the long-run, $\Delta\frac{W}{P} = \frac{\Delta TFP}{S_L}$ and $\Delta\frac{R}{P} = 0$. What does $\Delta\frac{K}{L}$ look like in the long run? We have to think about a point we have neglected thus far: technical bias. Consider Figure 19-1. As technology improves, the isoquant shifts inward, since we can produce the same amount as before using fewer inputs. If the isoquant gets steeper as it shifts inwards, this is a bias toward labor. There are three ways to think about this:

1. At a fixed $\frac{L}{K}$, $\frac{W}{R}$ is rising.

2. At a fixed $\frac{W}{R}$, $\frac{L}{K}$ is rising.

3. More *TFP* at higher *L/K*.

All of these say the same thing: we are twisting the isoquant clockwise as we shift it inward.

How do we measure technical bias? $\Delta B = \Delta W - \Delta R - \frac{1}{\sigma}(\Delta K - \Delta L)$. It is again a residual. This is the difference between $\Delta\frac{W}{R}$ and what we predict this change will be using the elasticity of substitution (that is, if there is no technological bias). If we want to measure it using quantities, we can write $\Delta B = \Delta L - \Delta K - \sigma(\Delta R - \Delta W)$.

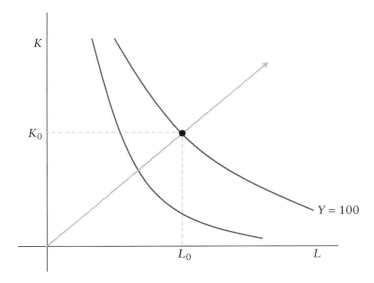

Figure 19-1: Technological bias toward labor.

From now on, we assume all the variables are measured in real terms (i.e., normalize the price to 1), so we will just write the change in the (log) real wage, for example, as ΔW. From last time, we related the change in total factor productivity to the (log) factor-price changes:

$$\Delta TFP = S_L \Delta W + S_K \Delta R.$$

To interpret this, remember that $P = MC = \dfrac{Factor\ Price}{Marginal\ Product}$. If marginal productivity rises 10%, this allows factor prices to rise 10% and maintain the same level of output price. We do not know which factor prices will rise, however; so we cannot say whether higher TFP will be seen in wages or capital prices. If labor and capital are fixed in the short run, the increase in output will be exactly the increase in productivity.

In the previous chapter, we also related factor-price changes to factor-quantity changes using the two-factor demand curves. We now consider the more general case where the bias is not zero:

$$\Delta W - \Delta R = \frac{1}{\sigma}(\Delta K - \Delta L) + \Delta B.$$

Some people will also talk about factor-augmenting technological progress, which is different from technological bias. In factor-augmenting technological progress, $Y = F(A_L(t)L(t), A_K(t)K(t))$, where A_L and A_K are labor-augmenting and capital-augmenting technological progress, respectively. This doesn't even tell us the direction of technological bias. With Cobb-Douglas production, for example, A_L and A_K are really the same thing, because we have

$$F(A_L(t)L(t), A_K(t)K(t)) = (A_L(t)L(t))^\alpha (A_K(t)K(t))^{1-\alpha}$$
$$= A_L(t)^\alpha A_K(t)^{1-\alpha} L(t)^\alpha K(t)^{1-\alpha}.$$

In general, the elasticity of substitution will determine which way A_L and A_K bias technological growth. If the elasticity of substitution is 1, they're both neutral.

Remember we can also measure technological progress from the quantities:

$$\Delta TFP = \Delta Y - (S_L \Delta L + S_K \Delta K).$$

In order to measure technological progress, we don't need to know much about the production function other than shares. Measuring technological bias is a different story, because it requires us to say something about σ, the elasticity of factor substitution in production. The bias can be signed, for example, if we know $\Delta W - \Delta R < 0$ and $\Delta K - \Delta L > 0$. In general, however, we will have to know the elasticity of substitution to measure technological bias.

RELATING LABOR'S SHARE TO ECONOMIC GROWTH

We have worked out three equations relating the percentage changes in different quantities and prices. These tell us the theoretically possible relationships between quantity changes, factor price changes, and technology changes. We apply them by taking special cases for factor supply, such as constant factor quantities, or a constant capital-rental rate, or Malthusian supply (a constant wage rate). This is the same supply and demand framework from the previous chapter, except that now technological change does not have to increase the capital demand curve by the same percentage that it increases the labor demand curve.

Returning to our short-run analysis (constant factor quantities), the factor-demand curves and TFP definition are:

$$\Delta W - \Delta R = \Delta B \rightarrow \Delta R = \Delta W - \Delta B$$
$$\Delta TFP = S_L \Delta W + S_K \Delta R.$$

These imply that

$$\Delta TFP = S_L \Delta W + S_K (\Delta W - \Delta B),$$

which simplifies to

$$\Delta W = \Delta TFP + S_K \Delta B.$$

So the change in the real wage, in the short run, will be affected by the technological progress as well as the bias of the progress. Remember that ΔB is defined as bias in favor of labor, so positive bias toward labor will mean the change in the wage is higher than it would be without the bias. What about the relative shares of labor and capital, $\dfrac{S_L}{S_K}$? By definition of shares,

$$\Delta \frac{S_L}{S_K} = (\Delta W + \Delta L) - (\Delta K + \Delta R),$$

where we continue to use Δ to denote change in logs. In the short run, we have assumed $\Delta L = \Delta K = 0$, so

$$\Delta \frac{S_L}{S_K} = \Delta \frac{W}{R}.$$

In general, the shares evolve according to

$$\Delta \frac{S_L}{S_K} = (\Delta W - \Delta R) - (\Delta K - \Delta L) = \left(\frac{1}{\sigma} - 1 \right)(\Delta K - \Delta L) + \Delta B,$$

where the second equality comes from the factor-demand curves. Factor shares were constant for a hundred years or so. During that "traditional" time frame, our model would be $\Delta S_L = 0$ and $\Delta R = 0$, which imply that $\Delta K = \Delta Y$ and $\Delta W = \dfrac{\Delta TFP}{S_L}$.

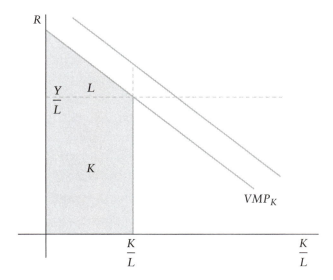

Figure 19-2: Short run versus long run. The total shaded area under the VMP schedule denotes Y/L. The part K is paid to capital, and the part L is paid to labor. In the short run, capital receives the gains of the additional productivity through a higher rental rate. In the long run, however, the rental rate will remain constant, and all of the additional surplus will accrue to labor.

To give more detail, consider the following example. We start in period 0, where $\frac{Y}{L} = 1$, $L = \alpha$, $K = 1 - \alpha$, $W = 1$, and $R = 1$. Suppose, in period 1, we know that $\frac{Y}{L}$ will be 10. Then the traditional model tells us that $W_1 = 10W_0$, $Y_1 = 10Y_0$, $R_1 = R_0$, $K_1 = 10K_0$, and $L_1 = L_0$. Labor share remains constant because the wage has gone up 10 times while the labor quantity has remained constant, and capital share has remained constant because the quantity of capital has gone up 10 times while the rental rate has remained constant.

But what if we saw an economy, still with $\Delta R = 0$, where labor's share was falling? If $\sigma = 1$, then it must be that $\Delta B < 0$, or that we have bias toward capital. But if $\sigma > 1$, there is another way to explain a reduction in labor's share even without biased technical change: increase in the capital stock.

Consider Figure 19-2. Before any shift, capital gets paid its marginal product, and labor gets paid the remaining surplus. These are depicted by K and L in the figure. If there is a zero-bias TFP increase then, in the short run, the demand curve shifts up and the rental rate rises; that is, capital gets the increase in the height of the curve. In the long run, however, the rental rate will remain the same, and all of the benefits from additional capital and increased productivity will accrue to labor. The additional income going to capital just offsets the opportunity costs of accumulating it.

When $\sigma > 1$, capital accumulation does not so easily drive the capital rental rate back to where it was. We get more capital accumulation than we would with $\sigma = 1$. Labor benefits from this extra capital. In other words, in the long run labor gets all of the productivity growth (the area of the quadrilateral bounded by the x-axis, the line at $\dfrac{K}{L}$, and the two capital-demand curves), plus some of the additional output created by the extra capital (this part of labor's benefit is the triangle to the right of the vertical line and above the dashed horizontal line).

The irony here is that with $\sigma > 1$, bias confers an extra benefit on labor at the same time that it reduces labor's share. Many people conclude that labor's falling share indicates harm to labor, but that conclusion is premature. It could be reflective of a capital-deepening process that redistributes from capital to labor.

Combining some of our equations, we have

$$\Delta W = \Delta TFP + S_K \left[\frac{1}{\sigma}(\Delta K - \Delta L) + \Delta B \right].$$

This decomposes the wage change into a TFP component, a capital-deepening component, and a bias component, regardless of whether we are talking about the short or long run.

Consider a world $\Delta B < 0$ and with $\sigma > 1$, so that technological progress is biased in favor of capital and capital substitutes more than one-for-one with labor. Looking back at the elasticity of substitution equation,

$$\Delta W - \Delta R = \frac{1}{\sigma}(\Delta K - \Delta L) + \Delta B.$$

So if $\sigma > 1$, in order for wages to go up more than capital-rental rates—as they must in the long run—we will need capital deepening or a bias toward labor. If there's a bias in favor of capital to begin with, then we'll need even more capital deepening. There will be an elevated capital share in the short run and an even larger increase in the capital share in the long run. But is this bad for workers? No. The same process driving up capital share is driving up wages and driving down rental rates.

THE MALTHUSIAN SPECIAL CASE

Note that Malthus's model is a special case of our change equations. Instead of the long-run fact that $\Delta R = 0$, however, he believed that $\Delta W = 0$. Because during his time, capital—largely land—was relatively inelastic, and labor supply (i.e., population) was very responsive to wage changes, he believed people would have more children in response to higher wages until the wage was driven back down to subsistence. In the neoclassical growth model, we have the exact opposite fact that capital is perfectly elastically supplied in the long run, so that $\Delta R = 0$.

Two things changed after Malthus. Capital changed, in that land is no longer as important a component of capital. The accumulation of people also changed; now, when people have more money they generally invest more in each child rather than having more children.

CAPITAL-BIASED TECHNICAL CHANGE ALSO BENEFITS LABOR

Let's look at an extreme case, where technical progress is biased rather extremely in favor of capital. In the worst-case scenario, where a worker owns no capital at all, will that worker still be better off? The answer is no, in the short run, but yes, in the long run. Consider Figure 19-3. In the short run, the rise in productivity yields much higher rents for capital owners, and the amount paid to workers is decreased (the triangle for labor has the same base in the short run, but due to the bias, a shorter height). In the long run, however, workers gain significantly more surplus than they used to have.

As with a lot of share phenomena in economics—consumer expenditure shares are another example—we need to know whether prices are increasing the share or some quantity change is doing it.

Now we'll suppose a case that might appear more favorable to workers. Suppose we have extreme technological bias toward workers. Consider Figure 19-4. Now, workers gain all of the surplus in the short run, but there is no capital-deepening response that would give them additional surplus in the long run.

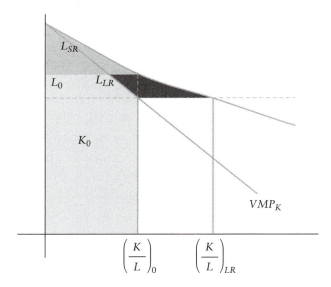

Figure 19-3: In period 0, labor earns the triangle labeled L_0 formed by the dashed horizontal line, the VMP_K line, and the y-axis. When productivity is increased, the surplus accruing to labor shrinks in the short run and is given by the area colored by the middle shade of gray and labeled L_{SR}. In the long run, the full triangle labeled by L_{LR} accrues to labor.

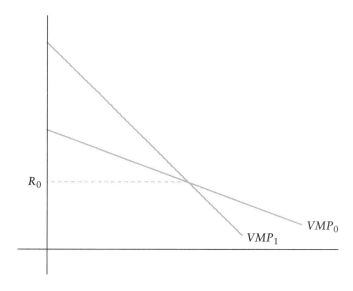

Figure 19-4: Technological bias in favor of workers. Here, there will be no capital deepening.

ADDING HUMAN CAPITAL

Now we are going to add human capital to this model. We will have three inputs: L_H, high-skilled labor; L_L, low-skilled labor; and K, physical capital; the three associated prices are W_H, W_L, and R, respectively. We want to consider the process of long-run growth. For this, we will want to consider (1) technological growth (TFP increasing), (2) capital deepening $\left(\dfrac{K}{L} \text{ increasing} \right)$, and (3) human capital deepening $\left(\dfrac{L_H}{L_L} \text{ increasing} \right)$.

Note there are several dimensions to L_H. It is useful to think about $L_H = N_H H_H t_H$, or the idea that the amount of high-skilled labor is the number of high-skilled workers multiplied by their human capital and the amount of time they spend working. Defining L_L similarly, we get that

$$\frac{L_H}{L_L} = \left(\frac{N_H}{N_L} \right)\left(\frac{H_H}{H_L} \right)\left(\frac{t_H}{t_L} \right).$$

The first term is the extensive margin of human capital investment; that is, we look at entry into high-skilled labor versus low-skilled labor. The second term is the intensive margin of human capital investment. We ask: within categories, how much human capital do the workers have?

We will assume technology is biased toward L_H relative to L_L, and that K is a substitute for L_L and a complement for L_H. As capital grows, it raises the relative demand for high-skilled labor and lowers the relative demand for low-skilled labor.

		Growth $\left(\dfrac{Y}{L} \right)$	$\dfrac{W_H}{W_L}$
(1)	Technological Growth	$+$	$+$
(2)	$\dfrac{K}{L}$	$+$	$+$
(3)	$\dfrac{L_H}{L_L}$	$+$	$-$

The negative relationship between $\dfrac{L_H}{L_L}$ and $\dfrac{W_H}{W_L}$ results because increasing skilled labor relative to unskilled labor will serve to drive the wages of the two groups closer to each other. Thus, inequality between workers is a race between the positive forces and the negative forces. The supply response of increasing $\dfrac{L_H}{L_L}$ serves to reduce inequality, but technological growth and increases in $\dfrac{K}{L}$ tend to increase inequality. In other words, if the effects $(1) + (2) < (3)$, then $\dfrac{W_H}{W_L}$ is decreasing, and the opposite is also true.

Chapter 20

Investments in Health and the Value of a Statistical Life

Here we treat investment in health for its own sake. We all invest in health. We make judgments about what our diets should be, whether we should visit a doctor, what medication we should take, whether we should stop drinking (or watch how much drinking we're doing), and so on. Modern medicine involves drugs and doctors, but modern medicine also involves a tremendous amount of personal input. Historically, going back to the 18th century, people didn't know how to make themselves healthier, and doctors didn't know either. There wasn't much personal knowledge that could go into it; of course, now we know that personal choice is one of the major determinants of people's health. Questions such as which doctors people see, what drugs they take, do they smoke a lot, do they drink a lot, do they eat a lot of food rich in fat, are important.

In short, people are now aware that many decisions affect their health. Their decisions have become more important. The emphasis here is what people can do for themselves by making investments in their health. That doesn't mean that they do without modern medicine; that's going to be subsumed in what they do.

Health investments link up with the microeconomics of the statistical value of life (SVL, also known as the value of a statistical life, or VSL). There is growing macro literature on the economy-wide effects of improvement in health and life expectancy. There are interesting papers by David Meltzer (1992), Acemoglu and Johnson (2007), and others on macro aspects of health, but we focus on the microeconomics of individuals' decisions.

INVESTMENTS IN SELF-PROTECTION

The micro approach to health is based on a couple of points: (1) the statistical value of human life (SVL), (2) optimal investment in health, and (3) health as self-protection. Self-protection is defined as people taking actions to reduce the probability of bad things happening (Ehrlich and Becker 1972). It is an important concept because you're going to see that the results we get for health look very much like self-protection. With market insurance, you trade off marginal utility in different states. You would like to transfer resources into states where your marginal utility of income is high. For instance, you may want to take out fire insurance because income is worth more in the state where your house burns down than in other states. Self-protection is different and may be a complement to insurance, but it is not insurance. You're changing behavior in order to lower the probability of your bad states. When you take out fire insurance, that may even raise the probability of bad states, right? Fire insurance creates a moral hazard: so what if your house burns down? The company will pay you off. So self-protection is very different. If you don't have fire insurance, you want to lower the likelihood of having a fire because you would lose your house and belongings—therefore, you will self-protect.

As another example of the difference between insurance and self-protection, take the question of why parents don't insure their children (empirically, they don't). A good explanation is that parents love their children, but they will only insure them if their marginal utility of income is going to be higher when their children die than when they don't die. But what they will do is self-protect. Parents sacrifice their lives for their children. Parents would give up a lot to reduce the probability that a child will die; nevertheless, they won't take out insurance on them. Instead of it being a paradox, it's a difference between the reasons for insurance and the reasons for self-protection. We will see that distinction as we analyze health decisions.

We start with a two-period model, where there is some probability of surviving to the second period; you don't survive through a third period, that's certain. Call the probability S—the probability of surviving to period 2. S is going to be a function that could depend on years of schooling s, or expenditures on health h. We have the following utility function:

$$V = u(x_1, l_1) + S(s, h)\beta u(x_2, l_2) + (1 - S(s, h))\beta U_D$$

where x and l denote consumption and leisure, respectively, and U_D is the utility if you die. Now if you die there may be some big utility, the magnitude of which may depend on your beliefs about the afterlife, for example. But if we look at the two S terms we have $S(s, h)(U_L - U_D) + U_D$, where U_L is the utility when alive. So really what we're looking at is the difference in the utility between living and dying. Now if the utility of dying is independent of all the decisions we are making (like h and survival), we can just treat that as a constant and get rid of it; we normalize it to zero for the purposes of predicting what people do.

Hereafter, we suppress the dependence of survival on schooling s, so that h is the only variable changing the survival probability. We assume that $\dfrac{\partial S}{\partial h} = S' \geq 0,\ S'' \leq 0$, survival is concave—you get smaller and smaller improvements as you spend more on health.

With the normalization, we have a simple two-period utility function $V = U(x_1, l_1) + S(h)\beta U(x_2, l_2)$, where x denotes consumption, l denotes leisure, and S denotes the probability of surviving until the second period and is a function of health investment.

Assuming full annuity insurance—that expected consumption including the expected expenditure on health is equal to your expected income—we have the lifetime budget constraint:

$$x_1 + S\frac{x_2}{1+r} + g(h) = w_1(1 - l_1) + \frac{S}{1+r}w_2(1 - l_2),$$

where $g(h)$ gives the expenditure on some health-augmenting activity.

We have the consumption first-order conditions $U_{x_1} = \lambda$ and $S\beta U_{x_2} = \dfrac{\lambda S}{1+r}$. Because we assume full annuity insurance, the survival term S cancels and uncertainty disappears from the condition, even though the model includes uncertainty about dying.[1] The first-order conditions for leisure are $U_{l_1} = \lambda w_1$ and $\beta S U_{l_2} = \dfrac{\lambda S w_2}{1+r}$.[2]

Now, if we differentiate with respect to h and use the shorthand $U(2)$ to represent the level of utility in the second period, we get

$$S'\beta U(2) = \lambda\left(g'(h) + \frac{S'}{1+r}[x_2 - w_2(1 - l_2)]\right).$$ The left-hand side gives the

marginal benefit: investing more into health raises the probability of surviving, S' being positive, but that's going to occur in the second period, so you have to discount it by β. But then you're picking up utility *level* $U(2)$, which is absolutely crucial to understanding the literature on the value of a statistical life. It's utility levels you're picking up, not marginal utilities; that's where self-protection comes in. That's why it's different from market insurance, where the benefits are marginal utilities.

The right-hand side gives the marginal cost: the direct cost plus the increased cost of the annuity if period 2 consumption is greater than period 2 income.

We can think about buying h as buying self-protection. Remember, we are buying complete annuity insurance. But we still might want to engage in self-protection. Why? Self-protection maximizes your expected discounted wealth, whereas insurance smooths it out over different states of the world. Even though optimal savings means we do not want to transfer wealth between the two periods, we still want to increase the probability of getting the second period utility. The marginal benefit of increasing h is not 0. Using the consumption condition $\lambda = U_{x_1}$, another way to write the first-order condition for health is

$$\frac{S'\beta U(2)}{U_{x_1}} = \left(g'(h) + \frac{S'}{1+r}[x_2 - w_2(1-l_2)] \right).$$

There is another reason we might invest in health. This explanation comes from the difference between average and marginal utility, the concavity of the utility function. When you lower the risk of dying, you pick up average utility, whereas when you smooth over time using insurance, you pick up marginal utility. Let's make this more explicit. We will assume the utility function is homothetic in time and goods of degree γ, where $\gamma \leq 1$. Recall that homogeneity of degree γ implies that $U(2) = \frac{1}{\gamma}(U_{x_2} x_2 + U_{l_2} l_2)$. We can divide everything through by U_{x_2}. Before writing the result, we will want to analyze $\frac{U_{l_2}}{U_{x_2}}$, which will appear on the right-hand side. Note that we have $\beta U_{x_2} = \frac{\lambda}{1+r} = \frac{\beta U_{l_2}}{w_2}$, which implies $\frac{U_{l_2}}{U_{x_2}} = w_2$. That is, the marginal rate of substitution between goods in a given period really only depends on the cost of time versus goods (recall the cost of goods was normalized to 1). This doesn't depend on any discounting because we are only focusing on substitutions within periods. Now we can divide $U(2) = \frac{1}{\gamma}(U_{x_2}x_2 + U_{l_2}l_2)$ by U_{x_2} to get

$$\frac{U(2)}{U_{x_2}} = \frac{1}{\gamma}(x_2 + l_2 w_2)$$

so that

$$S'\beta\frac{1}{\gamma}(x_2 + l_2 w_2) = g'(h) + \frac{S'}{1+r}[x_2 - w_2(1 - l_2)]$$

On the left-hand side, we now have an adjusted measure of "full consumption"—namely, consumption of goods and consumption of leisure. It is weighted by the difference between marginal and average utility, $\frac{1}{\gamma}$. We haven't done anything to simplify the cost side. If we take the special case $\beta = \frac{1}{1+r}$, we can rewrite our expression as

$$\frac{S'}{1+r}\left(\frac{1}{\gamma} - 1\right)(x_2 + l_2 w_2) = g'(h) + \frac{S'}{1+r}(-w_2)$$

We can understand this in the following way. Suppose you chose h not to maximize expected discounted utility but rather to maximize expected discounted wealth. Then you would solve

$$\max W = w_1 + \beta S w_2 - g(h)$$

which yields first-order condition

$$\beta S' w_2 = g'(h)$$

Note that this is the same as the large condition above when $\gamma = 1$. Thus max V is equivalent to max W when $\gamma = 1$. If this is the case, why maximize expected discounted utility—just maximize expected discounted wealth. Under these conditions, all one is doing is self-protection. But this is not the whole problem! When we have concavity, $\gamma < 1$, we will have $\frac{1}{\gamma} - 1 > 0$, and then $g'(h) > \frac{S'}{1+r}w_2$. What does this mean? You're spending more on health than you otherwise would just for the purposes of maximizing wealth. You pick up extra benefit from the concavity of utility. Thus your health spending is mostly done for the purpose of self-protection when $\gamma = 1$ or is close to 1, but it is also aimed at picking up the difference between marginal and average utility.

It may seem like this analysis is just a lot of formal nonsense, but it gives you a theory for thinking about optimal investment into health. It tells you that the individual is very important because γ is an important input into this problem. When $\gamma < 1$, we've said there is a difference between marginal and average utility. We can also say that there is con-sumer surplus. It also tells you that utility levels are important.

Life expectancy has an interesting complementarity. If I take an action to increase my probability of surviving next year, it also affects the uncon-ditional probability of surviving into the following years. Suppose we have three periods now. Then we have

$$V = U(x_1, l_1) + s_2 \beta U(x_2, l_2) + s_3 \beta^2 U(x_3, l_3),$$

Note that s_2 and s_3 are unconditional probabilities. That is, s_3 gives the probability of surviving into period 3 from the present, not the proba-bility of surviving to period 3 conditional on surviving in period 2. We can write

$$s_3 = P(survive\ to\ 2nd\ period)\ P(survive\ to\ 3rd | survive\ to\ 2nd\ period).$$

This is a complementarity type problem. The higher my probability of surviving to the third period conditional on surviving to the second, the more incentive I have to increase my probability of surviving to the second.

THE VALUE OF A STATISTICAL LIFE

We are interested in how much people are willing to pay to avoid certain risks. Also, we are interested in how much people must be paid to take on certain risks. Early studies looked at risky jobs, where one could die on the job, and jobs that were not risky in this way. They then asked how much people who take the risky jobs get paid relative to people who take the less risky jobs. Then you can try to tease out a measure about what the marginal person is willing to pay to take on a higher risk of dying.

We have a theory for this from our first first-order condition for health. Suppose the additional risk of dying from a job is 1 in 10,000. We can change s' in our condition by 1/10000 and simply assess how much the other side of the expression would have to change in order to hold with equality. This would measure how much more wealth one would need to be just as well off with a higher probability of dying. This could also be thought about, of course, using the indirect utility function.

In the data, it has been observed that if the increase in the risk is 1/10000 per year, you would have to be paid an additional $500 each year. Some studies find $200 to $300, others find $700, so this is just an aver-

age. The price per change in probability $\left(\text{i.e.,} \dfrac{500}{1/10000} \right)$ is $5 million.

This is what people mean when they talk about the value of a statistical life. Thus, in the United States as of 2010, the value of a statistical life has been estimated to be between $3 million and $7 million. Note that this depends on wealth or income and γ. In developing countries, the value of a statistical life might be something like one-sixth of the value in the United States. But does this mean those people aren't worth as much? No, it depends on what metric you are using. Suppose the government is considering whether to implement policies aimed at reducing the number of road accidents. There will be more incentive to do this in the United States. In many cases, poor countries spend less to reduce the probability of death for their inhabitants. In poor countries, people value the consumption (e.g., food) that the government could provide with that money more than they value road safety. We are not talking about the ethical worth of different individuals, but society's relative values of these things, which will vary depending on individual incomes of a country. And it does vary, empirically. For instance, U.S. per-capita health spending is much higher than the per-capita spending in many other countries.

Note that we have just used a small change to estimate the value of a statistical life. We don't have good estimates of the effect of big changes in the probability of death. Suppose we consider an old person who has the option of taking a drug to increase life expectancy by six months, and it costs $100,000 per year to take the drug. Some people say: what's six months? But others are willing to spend a lot.

Now let's think about whether or not the $5 million number we presented for the value of a statistical life is reasonable. We'll consider a back-of-the-envelope calculation for young people. Assume typical income in the United States is $40,000 annually. Now we add to this $72,000 for the value of household time, which makes up the larger chunk of one's time, assuming a 40-hour work week. Thus we get a number like $110,000 as an annual figure for full income. But we also know that γ, as discussed above, has been estimated in the literature to be about ½. Thus, we have to multiply our figure by 2. This gives us $220,000.

Then we can consider discounting at rate .04; then $\frac{1}{r} = 25$. Multiplying our full income figure by 25 then yields $5.5 million.

What about a country like India, where average income is roughly one-sixth of that in the United States? Suppose γ in India is the same as γ in the United States. Their value of life should be roughly one-sixth of that in the United States—approximately $800,000. They are not going to be willing to spend as much as someone in the United States on reducing the probability of death (because the marginal value of the consumption they would have to give up is higher).

Consider the recent swine flu scare. Why might it be rational to err on the side of caution? What would be the cost of a pandemic like the flu in 1918 and 1919? Look up some numbers on that and try to estimate the cost to the world today of a pandemic equal in magnitude to that of 1918–19.

Homework Problems for Part III

*Technological Progress and Markets
for Durable Goods*

1) Assume that telephone services are produced with labor, L, and capital equipment, K, according to the constant-returns production function $G(L, K)$, which has no technical change over time. Capital equipment is produced by a competitive industry with constant returns and, due to technological progress, the real cost of producing capital equipment is declining at $g\%$ per year. The equipment also has physical depreciation of δ per year and the interest rate is fixed at r.

    a. How will the current price of telephone services depend on the current level of productivity in the capital goods industry?

    b. How will the current price of telephone services depend on the rate of improvement of technology, g?

    c. How will the price of telephone services change over time if the real wage is growing at $x\%$ per year?

    d. If both g and x are fixed over time, will the price of telephone services fall over time at a constant rate?

2) Consider an industry that can invest in capital, K, which depreciates at rate δ. Investment goods are perfectly elastically supplied at price q. Once investments are made they cannot be reversed. The capital stock can only decrease due to depreciation. Industry output is produced with capital, K, and labor, L, with constant returns. The demand for the industry's output is subject to variation over time. Demand is either high, in which case industry demand is completely

inelastic at Y_H, or low, in which case industry demand is perfectly inelastic at Y_L. Labor is available at a constant wage W. When demand is high in one period, it will remain high with probability $p > 1/2$ in the next period and switch to being low with probability $1 - p$. When demand is low, a similar scenario holds where demand remains low with probability p and returns to being high with probability $1 - p$.

a. What will the pattern of investments and capital stocks look like over time?

b. How will the price of output compare when demand is high versus when demand is low?

c. What will happen to prices over time during periods when demand is high?

d. What will happen to prices over time during periods when demand is low?

e. How will the marginal product of capital compare in high- and low-demand states? Why?

f. How will operating profits of the firms compare in high- and low-demand states?

3) Consider the market for livestock (such as cattle or hogs). The industry maintains a herd of animals. Each year some of the herd is slaughtered while the remainder is held over for breeding (and possibly future slaughter). Denote the size of the herd in period t by H_t. Each period we slaughter some animals, denoted by X_t, and retain $(H_t - X_t)$ for breeding. Next period, the herd size will be $(H_t - X_t)(1 + g)$, where g is the natural rate of population growth for the herd. The cost of maintaining the herd is C per animal held over.

a. When will it pay for a farmer to hold an animal over compared to slaughtering the animal currently?

b. Using your results from part (a), how must the prices of animals be related over time in a market equilibrium? Why?

c. Find the steady-state price and herd size for animals when the demand for animals at slaughter is constant over time. Is there always a steady-state price and herd size? Why or why not?

d. Now assume that demand is constant but that we start with a herd size that is smaller than the steady-state herd size. How will the size of the herd, the number of animals slaughtered, and prices change over time as we converge to the steady state?

e. Now assume that we are in a steady state with a constant herd size, price and slaughter. How would a temporary (one period) rise in demand affect the current slaughter and price? How would it affect future herd sizes, slaughters, and prices?

f. How would a permanent increase in the demand for animals at slaughter affect prices, herd size, and the number of animals slaughtered over time?

g. How would a permanent increase in C affect the time paths of herd size, animal prices, and slaughters?

4) Businesses are now investing in "artificial intelligence" (AI), which we take to be a form of capital investment that is a better substitute for labor than historical capital investments have been.

a. *True, False, or Uncertain*: The degree to which AI substitutes for labor is largely a question of engineering or computer science rather than economics.

b. What does price theory say about the long-run effect of AI on the level of real wages rates paid to human workers? What is its effect on (human) labor's share of GDP?

c. What about the short-run effects on labor's share and the level of wages?

5) In the market for light bulbs, manufacturers have a choice as to how long their bulbs last (i.e., number of hours h that they can be turned on before permanently burning out). The market inverse demand for bulbs is $D(n,h)$, where n is the number of bulbs sold. The average and marginal cost of manufacturing a bulb is $c(h)$, where c is an increasing and convex function.

a. What is the relationship between inverse demand function and the marginal willingness to pay function?

b. Assuming that the remaining lifetime on a bulb can be readily observed, how would the equilibrium price of a used bulb be related to the price of a new bulb?

c. Can we conclude that $\dfrac{\partial D(n,h)}{\partial h} > 0$? What if each consumer were buying, at most, one bulb?

d. If consumers found it costly to change bulbs, how would their cost be reflected in the inverse-demand function D?

e. Does the equilibrium durability of bulbs depend on whether the market is competitive versus monopolized?

f. Would a per-bulb tax on new light bulbs affect the equilibrium durability? Does that depend on whether the market is competitive or monopolized?

6) Young adults are investing in their human capital, both in formal schooling and after they graduate. There is a technology $f(h)$ that translates time spent investing in human capital into middle-age hourly wage rates, with $f'(h) > 0$ and $f''(h) < 0$. That function varies across people for a variety of reasons (e.g., the rate at which they learn).

a. Between two otherwise identical young adults, one has a financial inheritance from his parents and the other is borrowing to finance his investments. The interest rate charged on the loans increases with the amount borrowed. Do the two accumulate the same human capital? Which one's investments are more sensitive to changes in the tuition charged by schools?

b. Referring to the two people cited in part a), do you expect either or both of them to work while in formal schooling? Explain.

c. Aside from the technology f, and access to funding, are there other reasons that you think that young people make different human capital investments? What are they?

7) *True, False, or Uncertain*: Computer manufacturing tends to be less volatile than housing construction because the depreciation rate of housing is lower.

8) *True, False, or Uncertain*: A tax on the income from working will discourage people from going to college.

9) *True, False, or Uncertain*: When housing prices are above their long-run values and continue to rise, that is good evidence that home buyers and home builders do not have realistic expectations about the future.

10) *True, False, or Uncertain*: Physicians work long hours because they are burdened with their medical-school debts.

11) *True, False, or Uncertain*: An increase in both employment and wages among women, relative to men, indicates that the relative demand for female labor has increased.

12) *True, False, or Uncertain*: An increase in the corporate-income tax rate reduces wages less than the revenue it obtains for the public treasury.

Notes

1. Quoted from Ross B. Emmett's (2010, p. 2) introduction to his volume on the "Chicago School of Economics."
2. A "flipped classroom" is one where the students are exposed to background material or general methods at home (with a textbook or video lecture) and use their time with their instructor to practice applying the methods.
3. If "intuitive" is meant as an antonym to mathematical rigor, we disagree. This book is mathematically rigorous in the sense that results are derived from fully specified assumptions. Indeed, many of the deductive results in this book have been performed on computers, which have no ability to fill in missing implicit assumptions (the symbolic computation representation of many of them can be found at http://examples.economicreasoning.com/). But the course emphasizes practice at applications—knowing what tools are available and when to use which ones—not practice at derivations.
4. Here we assume that, absent the subsidy, there is a strictly positive amount of corn going to ethanol production. This assumption is visible in the chart in that the supply curve always intersects the flat part of overall demand.
5. For large subsidies, the comparison is ambiguous because a large amount of corn may be drawn into the ethanol market and therefore require additional government revenue to finance the subsidy program. See also Figure I-2c, where some of the subsidy is paid to marginal supply that receives a net benefit of strictly less than x.
6. To derive this formula, totally differentiate the equilibrium condition $D_E(P_C - x) + D_F(P_C) = S(P_C)$ and solve for dP_C / dx. Multiply both numerator and denominator by $P_C E / C$ and use the fact that $D'_E + D'_F = D'_C$.
7. As shown in Figures I-2 and I-3, the price impact itself ranges from 0 to 1.
8. This simple model abstracts from timing, uncertainty, and other factors. In the more general case, the market may induce some people to be on the task-indifference ray

because, at the time that they acquire skills, they do not know which task they will end up doing. But even in this case, it will not make sense for everyone to be near that ray: some of them can be confident that they will be doing a particular task and thereby specialize in it.

CHAPTER 1

1. It is also sometimes known as Cournot aggregation, in order to distinguish it from the aforementioned Engel aggregation, which can also be referred to as "adding up." Hereafter we use "adding up" only to refer to Cournot aggregation.
2. See also chapter 3 and Becker's (1962) demonstration of a law of demand that relies only on the household budget constraint.

CHAPTER 2

1. Recall, if $\alpha \in [0,1]$ and F is concave, then $F(\alpha X_1 + (1-\alpha)X_2) \geq \alpha F(X_1) + (1-\alpha)F(X_2)$.
2. Hicks, *Value and Capital*, 2nd ed. (Oxford: Clarendon Press, 1946).

CHAPTER 3

1. Recall that symmetry and adding up together imply homogeneity.

CHAPTER 8

1. Typically, buyers of lower-quality housing will earn surplus, since the price is decreasing by the higher-quality buyers' willingness to pay, which exceeds lower-quality buyers' willingness to pay.

CHAPTER 9

1. The answer might appear to be different when the training time comes out of leisure rather than work time. But this only changes the form of the cost—forgone leisure instead of forgone earnings. The amount of the cost is the same as long as leisure and work time are priced the same, as they are when time is allocated optimally (see Ghez and Becker (1975) and Heckman (1976) for models of this type).
2. The same could be said in the context of on-the-job training if wage rate referred to the ratio of earnings to hours worked or trained.
3. Becker did not often find it useful to distinguish among various types of separation such as quits, layoffs, and retirements.

CHAPTER 10

1. Gary Becker took the same approach to analyzing the public sector (Becker 1983; Becker 1985; Becker and Mulligan 2003): posit a public-policy production function that subsumes the details of political processes but nevertheless has interesting things to say about how public policy making interacts with the rest of the economy. However, Becker's approach has been resisted by much of the political economics literature where it is asserted that political details are essential for predicting policy outcomes (e.g., Myerson 1995).
2. Lakdawalla and Philipson (2006) analyze nonprofit businesses in this way.
3. With constant returns to scale, marginal cost is independent of output and the cost function is proportional to output.
4. In the video, Professor Murphy wrote $F(X, Y)$ but here we replaced F with f and Y with Z because F and Y are used earlier in the lecture for different purposes. An example of a generic objective function f would be firm profits, part of which is the production function F.
5. This simple model abstracts from timing, uncertainty and other factors. In the more general case, the market may induce some people to be on the task-indifference ray because, at the time that they acquire skills, they do not know which task they will end up doing. But even in this case, it will not make sense for everyone to be near that ray: some of them can be confident that they will be doing a particular task and thereby specialize in it.

CHAPTER 11

1. Recall from chapter 7 that industry demand is the sum of each consumer's demand.
2. See Mulligan et al. (2018, Section 2.2) for an example.
3. Alternatively, the Δ operator can denote the change in the natural log (e.g., $\Delta Q \equiv \ln Q_{2015} - \ln Q_{2014}$).
4. Note that, by using the rules of calculus, we are considering small changes. Larger changes are examined by accumulating small changes, as shown in chapter 4. The uppercase S's refer to shares of revenue, whereas lowercase s's refer to shares of costs. They are the same with constant returns ($C = PY$), otherwise the S's do not sum to 1 across factors, even though the s's do.
5. Hicks, *Value and Capital*, 1946.

CHAPTER 12

1. For further reading on this approach, see Becker and Murphy (2006).
2. The consumption of drugs has a negative externality, which we have not specified in detail. Perhaps some of that negative externality is to reduce productivity in the legal sector.
3. The choke point is the point where the demand curve intersects the vertical axis (that is, price has gotten so high that all demand is "choked off").

CHAPTER 13

1. The height of the choke point is the price at which demand is zero.
2. For the moment we assume that all buyers are identical so that we can show an individual's demand curve in the same picture with the monopolist's marginal revenue curve.
3. The economic "theory of the core" makes a similar argument about gains from trade between buyers and sellers. See especially Telser (2006, 2009).

CHAPTER 14

1. Another way to arrive at this condition is to view it as part of a solution to a linear simultaneous-equations system. The system is formed by totally differentiating the (constant-returns versions of the) equilibrium conditions for p, Y, and X_i with respect to w_j, holding constant the other factor prices, and simultaneously solving for $w_j \, dp/dw_j$, dY/dw_j, and dX_i/dw_j. Replace $\partial C/\partial w_i$ and $\partial C/\partial w_j$ with X_i and X_j, respectively.
2. Here we have used $C = PY$ (constant returns) and $X_i = \partial C/\partial w_i$ and $X_j = \partial C/\partial w_j$.
3. Hicks, *Value and Capital*, 1946.

CHAPTER 17

1. For example, $C'(t)$ refers to $C'(\delta K(t) + \dot{K}(t))$.

CHAPTER 18

1. The same issue arises in the utility context, which is why the Laspeyres and Paasche quantity indices give different answers as to the change in utility.
2. We can derive this using the identity $Y = TFP \, F(K, L)$. Taking logs and differentiating gives us

$$\Delta Y = \Delta TFP + \frac{F_K K}{F} \frac{dK}{K} + \frac{F_L L}{F} \frac{dL}{L} = \Delta TFP + S_K \Delta K + S_L \Delta L.$$

3. Post-fisc income refers to own income minus taxes plus subsidies. In this case, workers pay no taxes but receive the revenue from the capital income tax.
4. The increase in workers' post-fisc income can also be written as the sum of five terms: $S_L(\Delta W + \Delta L) + \tau S_K(\Delta \tau + \Delta R + \Delta K)$. The second term is assumed to be zero. Substituting with the equations from the text for ΔW, ΔR, and ΔK, we have

$$S_K \Delta(1 - \tau) + \tau S_K \left(\Delta \tau - \Delta(1 - \tau) + \sigma \frac{\Delta(1 - \tau)}{S_L} \right) = \tau S_K \sigma \frac{\Delta(1 - \tau)}{S_L} < 0.$$

5. This assumes $F(0, L) = 0$. The results are the same if we allow $F(0, L) \neq 0$, but then we would have to track that amount. Also note that the capital demand curve here holds labor and technology constant, and not output.

6. *R/P* reduces both the base and the height of the labor-income triangle shown in the figure.

CHAPTER 20

1. If we further assumed that $\beta = \dfrac{1}{1+r}$, which results from some models, then the consumption-smoothing condition would merely be $U_{x_1} = U_{x_2}$.

2. Again, if $\beta = \dfrac{1}{1+r}$ and $w_1 = w_2$, we get that $U_{l_1} = U_{l_2}$.

Bibliography

Acemoglu, Daron, and Simon Johnson. 2007. "Disease and Development: The Effect of Life Expectancy on Economic Growth." *Journal of Political Economy* 115 (6): 925–85.

Becker, Gary S. 1957. *The Economics of Discrimination*. Chicago: University of Chicago Press.

———. 1962. "Irrational Behavior and Economic Theory." *Journal of Political Economy* 70 (1): 1–13.

———. 1964. *Human Capital*. New York: Columbia University Press.

———. 1968. "Crime and Punishment: An Economic Approach." *Journal of Political Economy* 76: 169–217.

———. 1971. *The Economics of Discrimination*. 2. Chicago: University of Chicago Press.

———. 1985. "Human Capital, Effort, and the Sexual Division of Labor." *Journal of Labor Economics* 3: S33–S58.

———. 1993. *Human Capital: A Theoretical and Empirical Analysis, with Special Reference to Education*. Chicago: University of Chicago Press. http://proxy.uchicago .edu/login?url=http://search.ebscohost.com/login.aspx?direct=true&db=eoh&AN =0329129&site=ehost-live&scope=site.

Becker, Gary S., and Gilbert R. Ghez. 1975. *The Allocation of Time and Goods over the Life Cycle*. New York: Columbia University Press (for NBER).

Becker, Gary S., and Casey Mulligan. 2003. "Deadweight Costs and the Size of Government." *Journal of Law and Economics* 46 (2): 240–340. http://www.jstor.org /stable/10.1086/377114.

Becker, Gary S., and Kevin M. Murphy. 1988. "A Theory of Rational Addiction." *Journal of Political Economy* 96 (4): 675–700. http://www.journals.uchicago.edu/loi/jpe.

———. 1992. "The Division of Labor, Coordination Costs, and Knowledge." *Journal of Political Economy* 107 (4): 1137–60. http://www.jstor.org/stable/2118383.

———. 1993. "A Simple Theory of Advertising as a Good or Bad." *Quarterly Journal of Economics* 108 (4): 941–64. https://academic.oup.com/qje/issue.

———. 2003. *Social Economics*. Cambridge, MA: Harvard University Press.

Demsetz, Harold. 1993. "George J. Stigler: Midcentury Neoclassicist with a Passion to Quantify." *Journal of Political Economy* 101 (5): 793–808. http://www.jstor.org/stable/2138595.

Ehrlich, Isaac, and Gary S. Becker. 1972. "Market Insurance, Self-Insurance, and Self-Protection." *Journal of Political Economy* 80 (4): 623–48. http://www.journals.uchicago.edu/loi/jpe.

Emmett, Ross B., ed. 2010. *The Elgar Companion to the Chicago School of Economics*. Northampton, MA: Edward Elgar Publishing.

Friedman, Milton. 1966. "The Methodology of Positive Economics." In *Essays in Positive Economics*, 3–43. Chicago: University of Chicago Press.

Grossman, Michael. 1972. "On the Concept of Health Capital and the Demand for Health." *Journal of Political Economy* 80 (2): 223–55.

Heckman, James J. 1976. "A Life-Cycle Model of Earnings, Learning, and Consumption." *Journal of Political Economy* 84 (4): S11-44. http://www.journals.uchicago.edu/loi/jpe.

Hicks, Sir John. 1946. *Value and Capital*. 2nd ed. Oxford: Clarendon Press.

Katz, Lawrence F., and Kevin M. Murphy. 1992. "Changes in Relative Wages, 1963–1987: Supply and Demand Factors." *Quarterly Journal of Economics* 107 (1): 35–78.

Klein, Benjamin, and Kevin M. Murphy. 2008. "Exclusive Dealing Intensifies Competition for Distribution." *Antitrust Law Journal* 75 (2): 433–66. http://www.jstor.org/stable/27897584.

Klemperer, Paul. 2004. "Why Every Economist Should Learn Some Auction Theory." In *Auctions: Theory and Practice*, by Paul Klemperer, 75–100. Princeton, NJ: Princeton University Press.

Lakdawalla, Darius, and Tomas Philipson. 2006. "The Nonprofit Sector and Industry Performance." *Journal of Public Economics* 90 (8–9): 1681–98. http://www.sciencedirect.com/science/journal/00472727.

Marshall, Alfred. 1890. *Principles of Economics*. London: Macmillan.

Meltzer, David. 1992. "Mortality Decline, the Demographic Transition, and Economic Growth." PhD Dissertation, University of Chicago.

Mulligan, Casey B. 2012. *The Redistribution Recession: How Labor Market Distortions Contracted the Economy*. New York: Oxford University Press.

Mulligan, Casey B., Russell Bradford, James H. Davenport, Matthew England, and Zak Tonks. 2018. "Non-linear Real Arithmetic Benchmarks Derived from Automated Reasoning in Economics." *Proceedings of the 3rd International Workshop on Satisfiability Checking and Symbolic Computation.*

Mulligan, Casey B., and Yona Rubinstein. 2008. "Selection, Investment, and Women's Relative Wages over Time." *Quarterly Journal of Economics* 123 (3): 1061–110.

Murphy, Kevin M., and Gary S. Becker. 2006. "The Market for Illegal Goods: The Case of Drugs." *Journal of Political Economy* 114 (1): 38–60.

Murphy, Kevin M., and Robert H. Topel. 2006. "The Value of Health and Longevity." *Journal of Political Economy* 114 (5): 871–904.

Neal, Derek. 1995. "Industry-Specific Human Capital: Evidence from Displaced Workers." *Journal of Labor Economics* 13 (4): 653–77. http://www.jstor.org/stable/2535197.

Pashigian, B. Peter, and James K. Self. 2007. "Teaching Microeconomics in Wonderland." *Journal of Economic Education* 38 (1): 44–57.

Rosen, Sherwin. 1972. "Learning and Experience in the Labor Market." *Journal of Human Resources* 7 (3): 326–42.

———. 1986. "The Theory of Equalizing Differences." In *Handbook of Labor Economics, vol. I*, edited by Orley C. Ashenfelter and Richard Layard, 641–92. Amsterdam: North-Holland.

———. 1988. "The Value of Changes in Life Expectancy." *Journal of Risk and Uncertainty* 1: 285–304.

Smith, Adam. 1776/1904. *An Inquiry into the Nature and Causes of the Wealth of Nations*. Edited by Edwin Cannan. London: Methuen & Co. http://www.econlib.org/library/Smith/smWN.html.

Stigler, George J. 1972. "Monopolistic Competition in Retrospect." In *Readings in Industrial Economics*, by Charles K. Rowley, 131–44. London: Palgrave Macmillan.

Telser, Lester G. 2006. *The Core Theory in Economics*. New York: Routledge.

———. 2009. *The Core Theory in Economics: Problems and Solutions*. New York: Routledge.

Thaler, Richard H., and Cass R. Sunstein. 2008. *Nudge*. New Haven, CT: Yale University Press.

United States Census Bureau. 2017a. "Historical Census of Housing Tables." https://www.census.gov/hhes/www/housing/census/historic/units.html.

United States Census Bureau. 2017b. "New Residential Construction." https://www.census.gov/construction/nrc/index.html.

Viner, Jacob. 1930/2013. *Lectures in Economics 301*. New Brunswick, NJ: Transaction Publishers.

Weyl, E. Glen. 2018. "Price Theory." *Journal of Economic Literature*.

Index

Page numbers in italics refer to figures and tables.

Acemoglu, Daron, 199

acquired comparative advantage, 12–16, *15*, 121–25, *122, 123, 124*

addiction: effects of declaring addictive goods illegal or overinflating bad effects of, 70–71; short-run and long-run price effects on addictive behaviors, 69–72, *70, 71*; using consumption stocks to understand, 66–69

adding up, 212n1; constraint on elasticities, 27; Hicksian demand functions and, 37; for the Marshallian system, 41–42

adjustment-cost model of investment, 17, 175–76, 179

allocation of resources, labor productivity and, 181, 182

all-or-nothing demand curve, 140–41, *141*

annuity insurance, 201, 202

applications: acquired comparative advantage, 12–16, *15*; demand for gasoline and cars, 62–66, *63*; ethanol fuel subsidies, 6–12, *7, 8, 9, 11, 14*; production-possibility frontier, 106–9, *107, 108, 109*; use in teaching price theory, 4–6

applied factor supply and demand: factor-biased technological progress, factor shares, and the Malthusian economy, 189–98, *190, 193, 196*; technological progress and capital-income tax incidence, 180–88, *181, 185, 187*

auction model of price, 3

Becker, Gary: on acquisition of differences among workers, 110, 125; on addiction, 67; demand and budget constraint and, 42; development of price theory and, 2; price change and shift in purchasing opportunities and, 44; price theory course and, 2, 4; public-policy production function, 213n1

behavioral economics, 16

boundary conditions, in rent gradient model, 95, 96

budget constraint: adding up and, 41–42; cost minimization, indifference curves, and, 31, *31*; income elasticity and, 43; indifference curves and, *35*, 35–36; investment in self-protection and, 201; rent

budget constraint:(cont.)
  gradient model and, *95*; utility
  maximization and, 21–23, *23*
budget line, quantity index and, 49–50
buyers: indifference curves for, 59–60, *60,*
  140–42, *141, 142*; price measurement
  and behavior of, 11–12. *See also*
  consumers

capital: adjustment cost model and,
  175–76; as complement or substitute,
  133–34; consumption-based valuation
  of, 177; depreciation of, 155–56;
  discounted value of net return on, 174;
  doubling output per unit of, 177–178,
  178; effect of corporate-income tax on
  corporate and noncorporate, 186;
  elastically supplied in the long run,
  186; examples of, 155; investment and,
  155, 156; land as, 155; law of motion
  for, 157, 168; in Malthusian economy,
  195; net return on, 174; output as
  function of labor and, 116; perturbations in steady state and, 159–164, *161,
  162, 163*; production of, 157; short-
  and long-run benefits from increased,
  193, 193–94; steady-state, 159, *160*
capital accumulation in continuous time,
  166–71; continuous-time versions of
  four equilibrium conditions and,
  168–71, *169, 170*; perturbing the
  steady state and, 166–68, *167*
capital-biased technical change, 195, *196*
capital deepening, 194; adding human
  capital and, 197; labor productivity
  and, 181, 182; wages and, 194
capital demand diagram, 185, *185*
capital gains, rental price and, 156
capital goods, use and investment
  markets for, 157, *158*
capital-income tax, incidence of,
  184–185, *185*
capital price, 156, 157; investment based
  on, 157, *158*
capital stock, price and, 169–70

capital subsidy, effect on employment,
  133–34
cars, demand for gasoline and, 62–66, *63*
chained price indices, 51–55, *52, 53, 54,
  55, 56, 57*
Chicago economics tradition, 1–2
Chicago price theory, 1–2; measurement
  and, 11–12; microeconomics *vs.*, 2–4;
  types of homework/test problems, 73;
  use of applications in teaching, 4–6
choke point, 55, 213n3; prohibition and,
  137–38
cigarettes, consumption and price of, 58,
  58
Cobb-Douglas production function, 145,
  191
comparative advantage: acquired, 12–16,
  15, 121–25, *122, 123, 124*; production-
  possibility frontier and, 106–9, *107,
  108, 109*
competition, price theory and, 2
complementarity: consumption over time
  and, 69; economic growth in presence
  of, 180–82, *181*; price theory and, 2;
  social interactions research and, 68
complementary goods, 2
complements: cross-price terms and, 66;
  demand for gas and cars and, 66;
  increase in quality and, 58; labor and
  capital as, 133–34
concavity of cost function, 32–33, *33,*
  34–35, *35*
conditional factor demands, 113;
  equation relating unconditional
  factor demands and, 114–15
constant-elasticity demand function, 135
constant returns to scale (CRS), 145;
  industry model and, 145–46; multiple-
  factor industry model and, 147
consumer demand, production and, 63–64
consumers: choices of goods by, 29, *29*;
  heterogeneous firms and, 91–93, *92,
  93*; negotiated discount and, 142;
  price theory *vs.* microeconomics and,
  2–3. *See also* buyers

consumer theory, 16; consumer misinformation and "nudgeability" and, 60–61, *61*; indifference curves for buyers and, 59–60, *60*; measurement and, 48; substitution effects and, 21. *See also* price indices

consumption: cost of increasing investment and, 176; effect of technological change on, 178, *178*; investment and, 186; monetary and health costs and cigarette, 58, *58*; past consumption influencing current, 67; price change and change in, *35*, 35–36, *36*, 37; prohibition and, 138; prohibition on illegal drugs and cost of, 135–36; relation of future consumption to present, 179; in short run and long run, 118

consumption-based valuation of capital, 177

consumption stocks, understanding addiction using, 66–69

continuous-time versions of equilibrium conditions, 168–71

corporate-income tax, incidence of, 186–88, *187*

cost function, 30–34, *31*, *33*; concavity of, 32–33, *33*, 34–35; indifference curves and, *35*, *36*; industry model and, 127; multiple-factor industry model and, 148, 149; properties of the, 32–33, *33*; valuing quality change using, 55–58, *58*

cost minimization, 30–32, *31*, 112–14, *113*; law of demand and, 34–35; rent gradient model and, 96; Slutsky equation and, 38

cost of living, measuring change in, *52*, 52–54

cost of production, up-sloping supply curve and, 172

costs, adjustment, applied to net investment, 174–76

Cournot aggregation, 212n1

crime rate, rent gradient model and, 97, 98

cross-price effects, cost function and, 35

cross-price elasticity, 26; Hicksian, 36–37

cross-price terms, gas-demand equation and, 66

demand: budget constraint and, 42; effect of ethanol fuel subsidies on, 7–11, *8*, *9*, *11*, *14*; for gasoline and cars, 62–66, *63*; Hicks' generalized law of, 34–35, *35*; housing boom and bust and, *164*, 164–165; perturbing steady state with rise in, 160–164, *162*, *163*, *164*; prohibition and, 137; prohibition on illegal drugs and, 135–36, *136*; Slutsky equation and, 40–41; theory of, 25–29, *27*, *29*. *See also* elasticity of demand; industry model

demand curves: all-or-nothing, 140–41, *141*; approximating with linear demand curve, 83; linear, 80, 81, *82*, *83*; normal distribution, 79, *80*, 81. *See also* Marshallian demand curves

demand elasticity: convergence to steady state and, 167–68; Marshallian demand equations and, 25–26; price of tires and, 57

demand equations, Marshallian, 25–26

demand functions: Hicksian, 32, 36–37; indifference curves and, *35*, *36*

demand system: degrees of freedom in, 43–44; indifference curves and, *35*, 35–36; relating short-run demand curve to overall, 64–66

depreciation of capital, 155–56; housing prices and, 166–67

distance, rent gradient model and, 97

distribution function, market demand as, 79–81, *80*, *82*, *83*

drugs, illegal: legalization of, 136–37; revenue from, 135–36

durable goods markets: homework problems, 207–9; price theory and, 17

durable production factors, 155–65; four equilibrium conditions, 157–59; perturbing the steady state, 159–65,

durable production factors (cont.)
161, 162, 163, 164; price theory and,
2; steady state, 159, *160*; stocks and
flows for factor prices and quantities
and, 155–56; use and investment
markets for capital goods and, 157,
158

economic growth: in presence of
complementarity, 180–82, *181*;
relating labor's share to, 191–94, *193*
efficient markets hypothesis, 160
elasticity (ies): constraints on, 26–28;
cross-price, 26; Hicksian demand
functions and, 36–37; multiple-factor
industry model and, 147–48; own-
price, 26; price and elasticity of
quantity, 129–30
elasticity of demand: cost function and,
33–34; for general distribution, 81;
income, 26–29, *27*, 43; labor and
capital and, 133–34
elasticity of substitution: capital
deepening and, 183; industry model
and, 131–32; partial, 146; technological bias and, 191
elasticity of supply, convergence to
steady state and, 167
elasticity version of Slutsky equation,
40
employers, human capital acquired from
training administered by, 101–2
employment, effect of capital subsidy
on, 133–34
endogenous factor prices, multiple-
factor industry model and, 149
endogenous interest rates, 176–79
Engel aggregation, 26, 43, 212n1
Engel curve, *27, 27*
equilibrium: increasing capital over time
and, *173*, 173–74; in rent gradient
model, 95–96. *See also* market
equilibrium
equilibrium compensating differences,
94–100, *95, 97, 98, 99*

equilibrium conditions, 157–59;
continuous-time versions of, 168–71,
169, 170
equilibrium price, ethanol fuel subsidies
and corn, 7, 10
equilibrium product quality, 81–87
ethanol fuel subsidies, 3, 4, 6–12; as
market "multiplier," 6–11, *7, 8, 9, 11,
14*; price theory guiding measurement, 11–12
exclusive dealing, 142–44
expenditure, defined, 48
expenditure growth, Laspeyres and
Paasche decompositions of, 48–51,
49, 50
explicit investment model, on-the-job
investment and, 101–2, *102*
extended price theory, 2

factor-augmenting technological
progress, 191
factor-biased technological progress,
189–91, *190*
factor demand: in multiple-factor
industry model, 147; substitution and
scale effects on, 120–21
factor-demand curves, 192
factor prices: multiple-factor industry
model and endogenous, 149; total
factor productivity and changes in,
190
factor shares, relating labor's share to
economic growth, 191–94, *193*
factor supply and demand. *See* applied
factor supply and demand
fast-food chains, Coke-only or
Pepsi-only, 143
feedback effects, 118
firms: cost minimization and, 112–14,
113; industry model, 16–17; marginal
cost of quality and, 85; production
function and, 109–10; profit maximization and, 110–12, *112*; Slutsky
equation of, 114–15, *115*; theory of,
108. *See also* heterogeneous firms

firm-specific investment in human capital, 104–5

Fisher ideal index, 51, 54, 55

flipped classroom, 5, 211n2

flow of new investment, 156

free lunch, 16; learning by doing and, 102, *103*

game theory, microeconomics and, 3

gas-demand equation, 66

gasoline: demand for cars and, 62–66, *63*; ethanol fuel subsidies, 6–12, *7, 8, 9, 11, 14*

Giffen good, 27, 41, *41*, 47

goods: adding up constraint on demand for, 27; Giffen, 27, 41, *41*, 47; homogeneity constraint on demand for, 27–28; income elasticity of demand for, 26–27, *27*; individual consumer value of, 79, *80*; luxury *vs.* necessity, 26; normal or inferior, 28, *28*

government, ethanol fuel subsidies and, 6–12, *7, 8, 9*

grocery stores: bundling items together to offset average cost above marginal cost, 144; negotiated discounts and, 142

habit formation, price theory and, 2

health: first-order condition for, 202; investment in self-protection, 199–204

hedonic models, 96

heterogeneous firms: consumers and, 91–93, *92, 93*; price and quality and, 87–90, *88, 89, 90, 91*

Hicks, John, 34; Marshall's Law and, 133, 148

Hicks' generalized law of demand, 34–35, *35*

Hicksian demand, symmetry for, 42

Hicksian demand curve, measuring change in cost of living and, *53*, 53–54

Hicksian demand functions, 32; properties of, 36–37

Hicksian system, Slutsky equation and, 38–41

homework problems: market equilibrium, 150–52; prices and substitution effects, 73–75; technological progress and markets for durable goods, 207–9

homogeneity: demand system and, 42; Hicksian demand functions and, 37; income elasticity and, 43

homogeneity constraint on elasticities, 27–28

housing boom and bust, 164–65, 166

human capital, 197–98; acquired comparative advantage and, 12–16, *15*, 121–25, *122, 123, 124*; acquired from training administered by employers, 101–2; labor productivity and accumulation of, 181, *182*; types of, 104–5

human capital analysis, 2

human capital deepening, 197

income: capital-income tax and labor, 184–85; increase in quality with rise in, 83–84, *84, 86*; post-fisc, 184, 185, 214n3; rent gradient model and, 96–97, *97, 98*

income-constant changes, Slutsky equation and, 39

income effect: of price change, *44*, 44–47, *45, 46*; Slutsky equation and, 39, 40

income elasticity of demand, 26–29, *27*, 43

India, value of a statistical life in, 206

indifference, price and indifference to purchase of good, 79, *80*

indifference curves: acquired comparative advantage and, 14; for buyers, 59–60, *60*, 140–42, *141, 142*; consumer choices of goods and, 29, *29*; cost minimization, budget constraints, and, 31, *31*; demand system and, *35*, 35–36; intersection of budget constraint and, 23, *23*; price line and, 84–87, *85, 86, 87*; production-possibility frontier and,

indifference curves (cont.)
107, *108*; quality, heterogeneous firms, and consumer, 87–89, *88, 89*; quality, profit, and firm, 89–90, *90, 91*; quantity index and, 49–50; rent gradient model and, 95, *96, 98*; for sellers, 140–42, *141, 142*; worker, *123*, 124

industry elasticity of labor demand, 132–33

industry model, 16–17, 126–34; four ingredients of, 131–32; industry elasticity of labor demand, 132–33; labor and capital as complements or substitutes, 133–34; multiple-factor, 145–49; properties of, 126–28, *127, 128*; review of, 145–46; supply-demand perspective on industry behavior, 128–30, *129*

industry-specific investment in human capital, 104, 105

inferior factors, 115, *115*

inferior inputs, 120–21

information problems with addictive goods, 70–71

input-demand functions, 111

input prices, value of marginal products equal to, 116, *117*

inputs: doubling output and doubling, *127*; inferior, 120–21

interest rates: endogenous, 176–79; price and reduction of, 166–68, *167*

inverse demand function, 172

investment: adjustment costs applied to net, 174–76; of capital, 155, 156; consumption and, 186; flow of, 156; in health, 199–204; perturbations in steady state and, 159–63, *162, 163*; from planning perspective, 172–79, *173, 177, 178*

investment-goods market equilibrium, 157; continuous-time version of, 168

investment market for capital goods, 157, *158*

isoprofit curves: heterogeneous firms and, 91–93, *92, 93*; monopolist's, 141, *142*

isoquant, bias toward labor and, *190*, 190–91

Johnson, Simon, 199

Klein, Benjamin, 143

labor: capital-biased technical change and, 195, *196*; as complement or substitute, 133–34; effect of corporate-income tax on, *187*, 188; effects of prohibition on, 137; in Malthusian economy, 195; output as function of capital and, 116; relating labor's share to economic growth, 191–94, *193*; in short run and long run, 116–17, *117, 119*; technological bias toward, 189, *190*. *See also* human capital

labor demand, industry elasticity of, 132–33

labor productivity: capital-income tax and, 184; definitions of, 180; growth in, 180–82, *181*

Laspeyres and Paasche decompositions of expenditure growth, 48–51, *49, 50*

Laspeyres price index, *50, 50*–51, 52; Fisher index and, 54, 55

law of motion for capital, 157; continuous-time version of, 168

learning by doing, on-the-job investment and, 101–3, *103*

legalization multiplier, 136–37

leisure: consumption of, 201, 203; learning by doing and price of, 212n1

leisure time, location choice and, 94

life expectancy, investment in health and, 204

linear demand curve, 80, 81, *82, 83*

linear simultaneous-equations system, 214n1

location choice, 94–100, *95, 97, 98, 99*

long run: benefits from increased capital in the, *193*, 193–94; capital elastically supplied in the, 186; consumption and the, 118; labor and the, 116–17, *119*

long-run demand: addiction and, 66–69; for cars and gasoline, 62–66, *63*

long-run price effects, on addictive behaviors, 69–72, *70, 71*

luxury goods, 26

maintenance model, 159

Malthusian economy, 195

marginal benefit and cost equation, 23–24

marginal cost: below average cost, *143*, 144; defining, 114; increase in output and decrease in, 115, *115*; investment in self-protection and, 202; output, factor demand, and, 120–21; output level equal to, 112–13, *113*

marginal cost curve, optimal output determined by consumer demand and, 128, *128*

marginal cost of production in production-possibility frontier, 106, *107*, 108, *108*

marginal products, input prices and value of, 116, *117*

marginal social benefit, of prohibition, 138–39

marginal utility: consumption over time and, 68, 69; investment in self-protection and, 202–4; prices and, 23–24

market analysis, price theory and, 1–2, 3–4

market demand, as distribution function, 79–81, *80, 82, 83*

market equilibrium: homework problems, 150–52; investment-goods, 157, 168; learning by doing and, 102–3; price theory and, 3; rental, 157, 168

market insurance, 200

markets, 16; effect of ethanol fuel subsidies on corn, 6–12, *7, 8, 9, 11, 14*; price theory and, 2–4

Marshall, Alfred, 1

Marshallian demand, unconditional factor demands *vs.*, 113

Marshallian demand curves, 16, 47; chained price index and, 53–54, *54*; market outcomes of the, 142–44

Marshallian demand equations, 25–26

Marshallian demand functions, 30

Marshallian system: adding up and symmetry for, 41–42; Slutsky equation and, 38–41

Marshall's Law, 132–33, *134*, 148

materials, 155

maximization, production function and, 109–10

maximization problem, 172–73

measurement: consumer theory and, 48; price theory and, 4, 11–12

Meltzer, David, 199

microeconomics, price theory *vs.*, 2–4

"monkey" solution, 33, *33*

monopolies, natural, 143–44

monopolists: isoprofit curves, 141, *142*; negotiated discount and, 142–44

monopoly models, price theory and, 2

multiple-factor industry model, 145–49; analyzing production and, 148; endogenous factor prices and, 149; properties of, 146–48

Murphy, Kevin M., 2, 67, 143

natural monopolies, 143–44

Neal, Derek, 104

necessity goods, 26

negative externality, of illegal drugs, 135, 137, 213n2

negotiated discount, 142

neoclassical growth model, 17, 179; long-run capital-supply curve for, 186; technological growth and capital deepening in, 182–83 nondurable goods, rise in demand and return to steady state, 161–62

normal distribution of demand, 79, *80*, 81

on-the-job investment: explicit investment model, 101–2, *102*; learning by doing and, 101–3, *103*

opportunity curves, specialization and, *124*, 124–25

opportunity set, for selecting human capital, 122–24, *123*

output: consumer demand, constant marginal cost curve, and optimal, 128, *128*; doubling inputs and doubling, *127*; doubling per unit of capital, 177–78, *178*; as function of labor and capital, 116; increase in price and increase in, 115, *115*; marginal cost and level of, 112–13, *113*

own-price elasticity, 26; multiple-factor industry model and, 147–48

Paasche price index, *50*, 51, 52; Fisher index and, 54, 55

pandemic, value of a statistical life and, 206

partial elasticity of substitution, 146

perpetual-inventory formula, 67

phase diagram, steady state, *169*

post-fisc income, 214n3; effect of capital-income tax on, 184, 185

preferences, theory of, 23

price control, 3, 150

price discrimination, ethanol fuel subsidies and, 10

price-index problem, 12

price indices, 48–58; chained, 51–55, *52, 53, 54, 55, 56, 57*; Laspeyres and Paasche decompositions of expenditure growth, 48–51, *49, 50*; using cost function to value quality change, 55–58, *58*

price-quantity relationship, 81, *83*

price(s): auction model of, 3; capital, 156, *157*; change in quantity and price between years, 129, *129*; determined by supply side, 128; elasticity of quantity and, 129–30; homework problems, 73–75; housing boom and bust and, 164–66, *166*; income effect of change in, *44, 44–47, 45, 46*; marginal utility and, 23; perturbations in steady state and, 160–65, *162, 163*; quality and, 55–58, *84, 84–87, 85, 86, 87*; rental, 116, 156; setting equal to marginal cost, *112, 113*; substitution effects and, 16

price-theoretic perspective on the core, 140–44, *141, 142, 143*

price theory. *See* Chicago price theory

price-utility vectors, Hick's generalized law of demand and, 34

production: consumer demand and, 63–64; marginal cost of quality and, 85; multiple-factor industry model and analysis of, 148; two-input, 116–20, *117, 119*; types of inputs, 155

production cost, prohibition and, 138, *138*

production function, 109–10; Cobb-Douglas, 145, 191; cost minimization and, 112; first-order conditions for, 148; multiple-factor industry model and, 148, 149; public-policy, 213n1

production-possibility frontier, 176, *177*; comparative advantage and, 106–9, *107, 108, 109*

product quality: equilibrium, 81–87, *84, 85, 86, 87*; price theory and, 2

profit maximization, 110–12, *112*

profits, defined, 110

prohibition, consequences of, 135–39, *136*; cost of half-hearted prohibitions, 137–39, *138*; legalization multiplier and, 136–37; revenue from drug sales and, 135–36

public-policy production function, 213n1

purchase-price equation, 157; continuous-time version of, 168

quality change, using the cost function to value, 55–58, 58
quality of housing, rent gradient model and, 97–99, 99
quality-quantity problem, 83–84, 84
quantity: change in price and quantity between years, 129, 129; price and elasticity of, 129–30
quantity discounts, 142–44
quantity index, expenditure growth decomposed into, 48–51
quasilinear utility, 82

rational expectations, market efficiency and, 164
rehabilitation for addictive behaviors, 70–71
rental market equilibrium, 157; continuous-time version of, 168
rental price, 156; capital gains and, 156; of labor, 116
rental rate, steady-state, 159, 160
rent gradient model, 94–96, 95; properties of, 96–100, 97, 98, 99
rents: effect of rise in demand on, 161–63, 162, 163; housing boom and bust and, 164–65; stocks and return to steady state, 170–71
returns to case, industry model and constant, 131
returns to scale, industry model and constant, 126, 127–28
revenue from illegal drug sales, 135–36
risk, value of a statistical life and taking on, 204–5
Rosen, Sherwin, 16, 103

saddle path, 169, 169–70, 170
scale effects: on factor demand, 120–21; industry elasticity of labor demand and, 132, 134; multiple-factor industry model and, 147

self-protection: defined, 200; insurance vs., 200; investments in, 200–4
sellers: indifference curves for, 140–42, 141, 142; price measurement and behavior of, 11–12
short run: benefits from increased capital in the, 193; consumption and the, 117–18; labor and the, 116–17, 117
short-run demand: addiction and, 66–69; for cars and gasoline, 62–66, 63; relating the short-run demand curve to overall demand system, 64–66
short-run elasticity of labor demand, 133
short-run price effects on addictive behaviors, 69–72, 70, 71
Slutsky correspondence, 38–39, 39
Slutsky equation, 38–41; demand system and, 42; elasticity version of, 40; of the firm, 114–15, 115; income effect and aggregating the, 46–47
Slutsky equation for the firm, substitution and scale effects and, 119, 120–21
Smith, Adam, 16
social cost, of corporate-income tax, 187
social interactions: complementarity and, 68; price theory and, 2
Sonnenschein, Hugo, 45
specialization: comparative advantage and, 12, 14, 15; firm-specific investment, 104–5; human capital, 124, 124–25; industry-specific investment, 104, 105
statistical value of life (SVL). See value of a statistical life (VSL)
steady state, 159, 160; capital accumulation in continuous time and perturbing the, 166–68, 167; perturbing the, 159–65, 161, 162, 163, 164; speed of convergence to, 166–68, 170–71
steady-state consumption, short- and long-run price effects and, 69–70, 70, 71
stock of durable assets, 156

structured problems, Chicago price theory and, 73

substitutes: cross-price terms and, 66; increase in quality and, 58; labor and capital as, 133–34

substitution: income effects and, 45; industry elasticity of labor demand and, 132; partial elasticity of, 146. *See also* elasticity of substitution

substitution effects, 16; consumer theory and, 21; on factor demand, 120–21; homework problems, 73–75; multiple-factor industry model and, 147; Slutsky equation and, 39–40

substitution equation, industry model and, 146

supply, short-run demand curve and, 65–66

supply curve, upward-sloping, 172

supply-demand perspective on industry behavior, 128–30, *129*

supply function, 111; multiple-factor industry model and, 149

supply side, determination of price and, 128

surplus, maximizing, 172–73

symmetry: elasticities and, 36–37; for Hicksian demand, 42; income elasticities and, 43; for the Marshallian system, 41–42; relation to adding up and homogeneity, 37

task-indifference ray, acquired comparative advantage and, 121–22, *122*

taxes: capital-income, 184–85, *185*; corporate-income, 186–88, *187*; price theory and, 4

technical change, capital-biased, 195–96, *196*; labor and, 195, *196*

technological bias: definition of, 189–91, *190*; in favor of workers, 195, *196*; measuring, 189

technological change: consequences of unbiased, 182–83; effect on consumption, 178, *178*

technological progress/growth: adding human capital and, 197; homework problems, 207–9; labor productivity and, 180, 181, 182; measuring, 191

TFU (true, false, uncertain) problems, Chicago price theory and, 73

theory of the core, 214n3

three-good model, 43–44

time constraints, utility maximization and, 21–22, 24–25

tires, demand elasticity and price of, 57

total factor productivity (TFP), 192, 193–94; capital-income tax and, 184; factor-price changes and change in, 190; price-based measure of, 183

trade: comparative advantage and, 12; indifference curves for buyers and sellers and, 140–42, *141, 142*; production-possibilities frontier and, 107–9, *109*

transaction price, price theory and, 3

travel time model, 94

turnover, firm-specific investment and, 105

two-good model, 43

two-input production, 116–20, *117, 119*

unconditional factor demands, equation relating conditional factor demands and, 114–15

uniform distribution, 80, *82*

United States, value of a statistical life in, 206

urban density, rent gradient model and, 100

use market for capital goods, 157, *158*

utility, difference between living and dying and, 200–1, 202–4

utility-constant changes, Slutsky equation and, 39

utility function, product quality and, 82

utility maximization, 21–25, *23*; demand equations and, 25–26; profit maximization *vs.*, 110–11

value of a statistical life (VSL), 199, 200, 204–6
Viner, Jacob, 1

wage inequality, rent gradient model and, 96–97, *97, 98*
wages: capital deepening and, 194; capital in the long run and decrease in, 118, *119*; effect of on-the-job investment on, 101–3, *102, 103*; firm-specific investment in human capital and, 104–5; industry-specific investment in human capital and, 104; labor in the short run and decrease in, 116–17, *117*; labor productivity and rising real, 180; setting equal to value of marginal product, *112*; technological progress and change in, 192; total factor productivity and, 183
Walmart, 28
willingness-to-pay-money schedule, 136
work time, price of, 44, 212n1